I. Kaplan · S. Giler

CO_2 Laser Surgery

With 322 Figures, mostly in colour

Springer-Verlag
Berlin Heidelberg NewYork Tokyo 1984

Isaac Kaplan,
Professor of Surgery, Head of Department of Plastic
and Maxillofacial Surgery

Shamai Giler
Senior Lecturer in Surgery

Beilinson Medical Center
Beilinson Hospital, Kupat-Holim-Health Insurance Institution
of the General Federation of Labour in Israel
Petah Tiqva, Israel
and the University of Tel Aviv Medical School

ISBN-13:978-3-642-82182-0   e-ISBN-13:978-3-642-82180-6
DOI: 10.1007/978-3-642-82180-6

Library of Congress Cataloging in Publication Data:
Kaplan, I. (Isaac), 1919–. CO_2 Laser surgery.
Bibliography: p. Includer index. 1. Lasers in surgery. 2. Carbon dioxide lasers –
Therapeutic use. 3. Skin-Surgery.
I. Giler, S. (Shamai), 1943. II. Title. III. Title: Carbon dioxide laser surgery.
RD73.L3K37 1984 617'.91 83-20183

2119/3020-543210

Preface

Over the past ten years, carbon dioxide laser surgery has made impressive strides and is now applied to every field of surgery without exception. It is the intention of this book to record the work done in this field in the Department of Plastic and Maxillofacial Surgery of the Beilinson Medical Center and Tel Aviv University Medical School, Israel, as well as that performed in association with other departments.

In this context, one feels that it is incumbent upon one to acknowledge the cooperation of the medical and paramedical staff of the Department of Plastic and Maxillofacial Surgery of the Beilinson Medical Center, as well as that of Prof. Yehuda Shindel and Dr. Daniel Katenelson of the Department of Ear, Nose and Throat, Dr. Yona Tadir of the Department of Obstetrics and Gynecology and Dr. Itamar Kott of the Department of General Surgery. I should like to make special mention of Dr. Ralph Ger of New York, who worked with me on the original clinical trials, and the engineer Uzi Sharon, who developed the Sharplan Laser with me.

The progress of Laser Surgery is well demonstrated by the participation in the four meetings of the International Society for Laser Surgery, the first of which was held in Tel Aviv in 1975 with an attendance of 65 and the last in Tokyo in 1981 with an attendance of 1200.

The motto of the society reads "Yehi Or" ("Let there be light") and I am sure that we can now add "Veyehi Or" ("And there was light").

In conclusion, I wish to express my deep appreciation to my wife Masha who accompanied me throughout and without whose moral support I could never have stood the pace. This book is therefore humbly dedicated to her.

October, 1983   Isaac Kaplan

Table of Contents

Part A

Carbon Dioxide Laser Surgery

With the development of suitable instrumentation, engineered to fulfil the requirements of the clinical surgeon and designed to meet the physical conditions prevailing in the operating theater, laser surgery has advanced rapidly in the past few years. There are signs of even more rapid progress and general acceptance as its application and advantages are demonstrated in more and more fields of surgery.

The basic advantages are as follows:

1. Noncontact surgery
2. Dry-field, almost bloodless surgery
3. Highly sterile surgery
4. Highly localized and precise microsurgery
5. Clear field of view and easy access in confined areas
6. Prompt healing with minimal postoperative swelling and scarring
7. Apparent reduction in postoperative pain
8. No electromagnetic interference on monitoring instrumentation

A laser surgical system consists basically of the laser – a source of intense visible or infrared radiation that can be focused to submillimeter-size-spots – and a delivery system that conveys this radiation to either a handpiece, a microscope attachment, or an endoscope.

1 The Carbon Dioxide Laser

Among the various systems currently used in medical practice, the CO_2 laser has proved itself to be the most efficient laser "scalpel" as a result of its infrared wavelength of 10.6 μm, which, unlike visible wavelengths, is highly absorbed in water. When the focused beam of this laser is incident upon living tissue (75%–90% water), the resulting effect is that of highly localized tissue removal through evaporation. The tissue-vaporized zone is surrounded by a thin layer of heat-coagulated tissue in which blood vessels smaller than 1 mm are sealed resulting in a dry, almost bloodless procedure.

During laser surgery, both beam position and the tissue undergoing surgery are under the continuous visual control of the surgeon. The progress of tissue incision is continuously monitored. Small volumes of tissue are removed with single, or multiple, exposure pulses, whereas larger incisions are performed through progressive scans of the beam in the continuous mode.

2 The Sharplan Systems

The apparatus employed by ourselves and others engaged in clinical surgery using this modality are the Sharplan systems. These are the only systems available for both free-hand surgery and microsurgery. Both systems employ flowing-gas and water-cooled laser heads. The laser medium is a mixture of CO_2, N_2, and He pumped through a water-cooled discharge tube at a pressure of 25 mm Hg.

The discharge tube is terminated by two mirrors, one fully reflective and one partially reflective, which constitute the optical resonator. The discharge energy is transferred via the N_2 molecules to the CO_2 molecules, which emit infrared radiation at a wavelength of 10.6 µm. This radiation is amplified in the optical resonator to produce the high-intensity collimated laser beam. The optical resonator is specially designed to assure the best optical-beam shape, designated TEM_{00}. This lowest beam mode is essential for the best surgical results because it is the only beam pattern with the following characteristics:

(1) Gaussian power distribution across the beam – no "hot spots" and peak power at beam center; (2) maximum focusing capability – no "diffraction-limited" spots; (3) minimal beam divergence – assuring highest efficiency in passage through delivery systems and attachments such as endoscopes.

The laser head is mounted horizontally on a telescopic column and houses, besides the CO_2 laser, a power detector, an electromechanical beam shutter, and a patented twin-beam He-Ne laser-aiming system. This system is designed to aid the surgeon in aiming the invisible CO_2 laser beam both prior to and during its application in free-hand surgery and microsurgery. The He-Ne laser emits a low-power (2-mW) visible red beam. The patented optical system splits this red beam and positions the resulting two beams on two sides of the main CO_2 laser beam. These beams are directed into a mirrored articulated arm that facilitates maneuverability of the beam in a radius of 2 m around the main cabinet.

This concept is utilized since optical fibers for 10.6-µm radiation (at the powers required for surgery) are not yet available.

2.1 Laser Attachments

For free-hand surgery, various penlike focusing handpieces can be attached to the articulated arm (Fig. 3a). These contain a lens that focuses the CO_2 laser beam and refracts the He-Ne laser beams to intersect at the precise focus of the CO_2 laser beam. The surgeon thus has a three-dimensional, continuously visible indication of the incision point. Standard focal lengths are 50 mm and 125 mm, with focal beam spot diameters of 0.1 mm and 0.23 mm, respectively. The focusing handpiece is also fitted with a mechanically operating guide, which can be removed for noncontact surgery. Upon activation of the main beam, dry nitrogen gas is passed through the handpiece to the incision area to retard oxidation and blow away fumes. It also serves to cool the lens.

For microsurgery, two different microscope attachments have been developed, which can be simply installed on standard operating microscopes requiring no modification of the microscope. These attachments are designed to reflect both the CO_2 laser beam and the aiming beam onto the microscope field of view and to facilitate maneuvering of the beam by the surgeon. In one version, this is accomplished through a patented mechanical micromanipulator with $\times 7$ angle magnification (Fig. 2a). In another version, electrically activated gimbals are used with a mirror. The beam in this version is controlled by a joystick placed on a remote control console. Both "Microslad" units are equipped with quick-change lenses (for 200-mm and 400-mm microscope objective lenses) and two-stage defocusing capability. Focal spot diameters of the CO_2 laser beam are 0.45 mm and 0.8 mm, respectively.

Additional attachments to the Sharplan system have been designed as custom items for specialized research programs. These include a laser cystoscope fitted with a Hopkins telescope and a ventilation bronchoscope; rectoscopes and laparoscopes are already in clinical use (Figs. 2c, 3c) as is a computerised scanner.

2.2 Simplicity of Design

The Sharplan system is designed for simple operation and control by the nontechnical surgeon. The various models available are shown in Figs. 1−4: It is ready for use within 20 s of being switched on. Once the appropriate power is set and monitored, the surgeon must choose the mode of operation, either continuous or pulsed. In the continuous mode, laser radiation is emitted for as long as the foot switch is pressed. In the pulse mode, each foot switch depression emits a preset "dosage" of radiation at pulse-lengths of 10 ms to 0.5 s.

Medical laser systems must comply with Bureau of Radiological Health (BRH) regulation (21 CFR 1040.10). The Sharplan incorporates all the required safety devices and, in addition, has a panel to indicate malfunction and even reminds the surgeon to open the gas cylinder.

Figure 1 a–d

(a) The Sharplan 791 was the first CO_2 laser designed specifically for clinical use. It has a maximum output of 50 W

(b) The Sharplan 733 was introduced into clinical use for soft-tissue surgery since its maximum output of 35 W was found to be adequate

(c) The Sharplan 743 with maximum output of 80 W and an optional rapid-pulse setting was introduced for orthopedic surgery and operations in which rapid vaporization or massive extirpation are desired

(d) The Sharplan 720 with a maximum output of 25 W was designed for private practice and ambulatory procedures in gynecology, dermatology, and ear, nose, and throat surgery

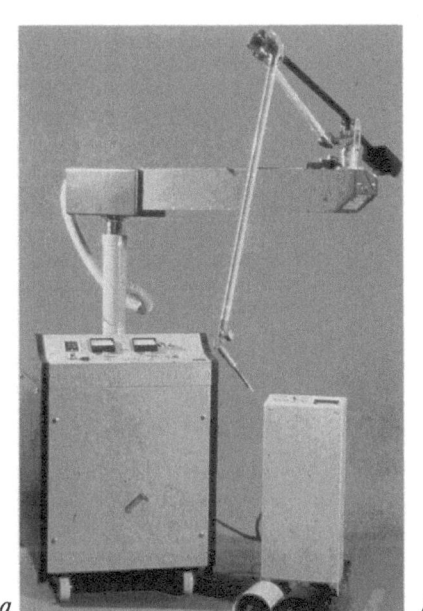

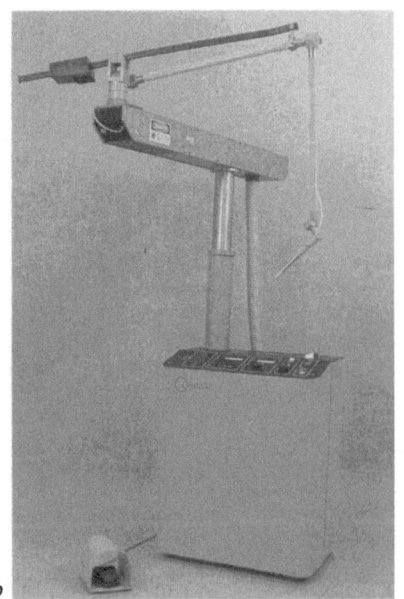

a

b

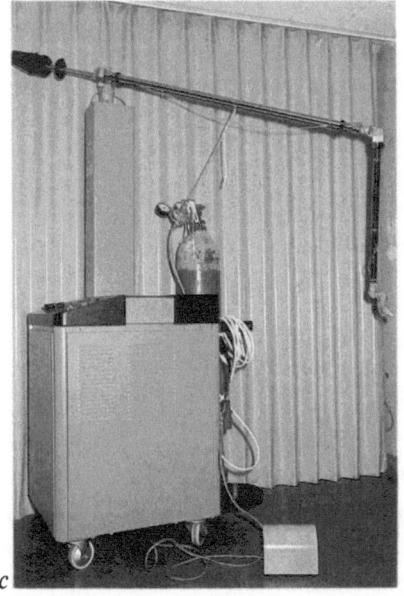

c

d

Figure 2a–c

(a) The micromanipulator attached to a Sharplan laser
(b) The electronically controlled micromanipulator designed for remote control in neurosurgery
(c) A prototype of a bronchoscope for transmitting the laser beam

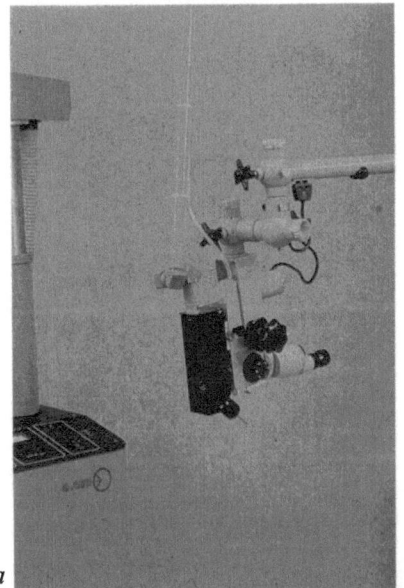
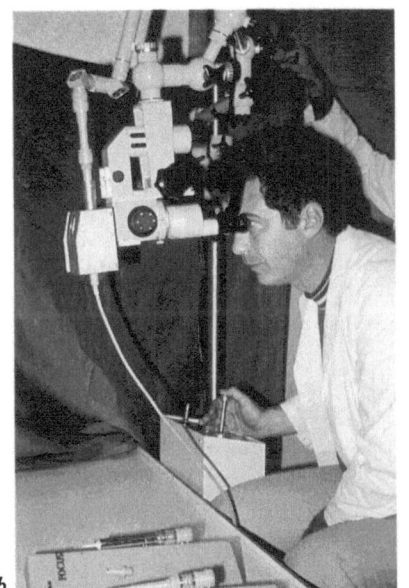

a

b

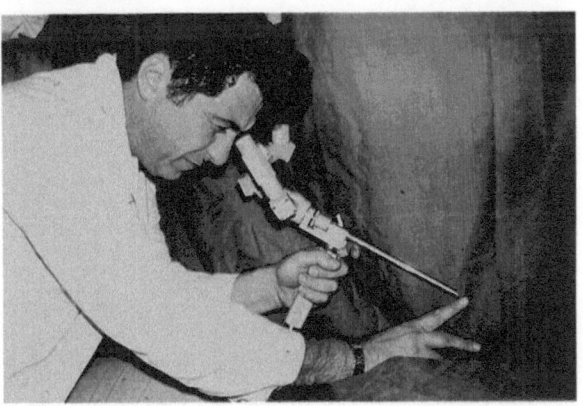

c

Figure 3 a–c

(a) Various focusing heads
(b) Laparoscopic attachment with a trocar for double-puncture use
(c) The Sharplan 720 attached to a colposcope

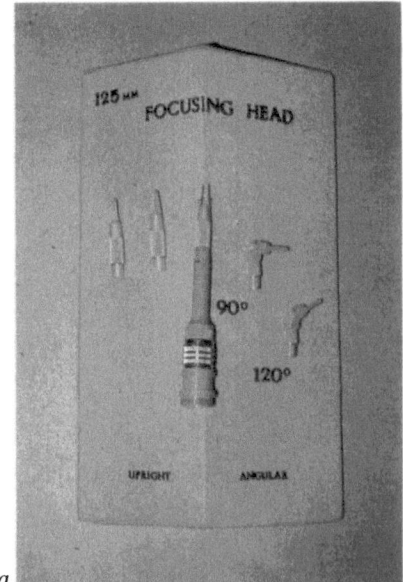

a

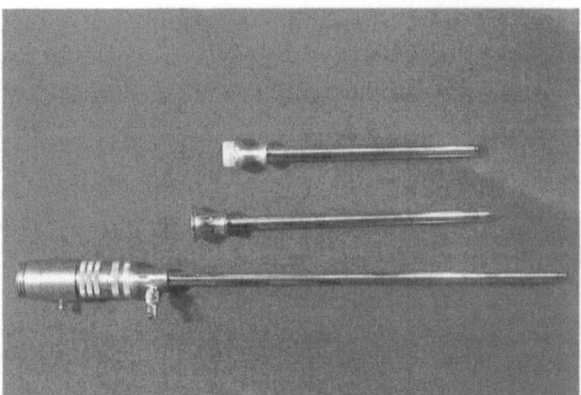

b

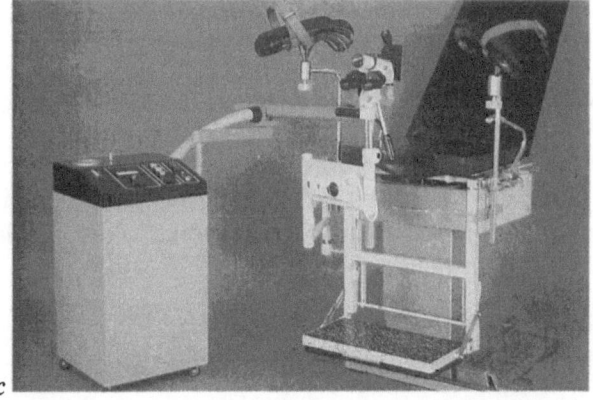

c

3 The Carbon Dioxide Laser in Practice

The Sharplan CO_2 surgical laser is now in routine use in more than 30 countries throughout the world, in every field of surgery without exception. We ourselves have performed over 3000 surgical procedures of various kinds and believe that certain definite statements can now be made as to the application of this modality in clinical surgery and its advantages over others.

3.1 Operations in Which the Anticipated Blood Loss is Significant

The risk of severe blood loss exists in practically every surgical specialty, but is particularly significant in orthopedic and plastic surgery, in which the excisions are mostly large. Mastectomies, mammaplasties, lipectomies, and other procedures of a similar nature have been performed by ourselves and others using the laser. In addition, orthopedic procedures, such as spine fusions and hip replacements, have been performed in many centers with statisfactory results.

It is worth mentioning that not only is the saving of blood impressive in these cases but the fact that the use of a tourniquet can be avoided reduces postoperative morbidity. Pediatric and neonatal surgery are particularly affected here because of the vital importance of saving blood in children.

3.2 Surgery Performed on Highly Vascular Areas of the Body

Perhaps the best examples of this are partial hepatectomies and partial nephrectomies. In our unit, tongue surgery and surgery of the scalp have been performed routinely by means of the CO_2 laser with impressive results. Experimental partial splenectomy and pancreatectomy show promise of the clinical application in these areas.

3.3 Extirpation of Highly Vascular Tumors

Cavernous hemangioma, Kaposi's sarcoma, and hemangiosarcoma have been extirpated by ourselves and others on many occasions. Impressive results in dealing with meningiomas and the advantage of the CO_2 laser in conservative myomectomies have been well established.

3.4 Surgery Performed on Patients with Bleeding Tendencies

Perhaps one of the most dramatic applications of the CO_2 laser lies in surgery performed on hemophiliac and thrombocytopenic patients. Major orthopedic procedures have been performed on patients suffering from hemophilia in whom there was not only a reduction in blood loss and postoperative morbidity but also a striking reduction in the expenditure involved in the preparation of antihemophilia factors. Our own experience in this connection in association with such institutions as the Wadley Institute in Dallas, Texas as well as that of others has confirmed these findings not only with regard to hemophilia and thrombocytopenia, but also in patients being treated with heparin and coumarin.

3.5 Surgery in Malignant Disease

It is universally accepted by surgeons that the surgery of cancer should be performed with the minimal opening of blood vessels and lymphatics, the minimal manipulation of tissue, together with maximal visualization. The CO_2 laser seals the blood vessels and lymphatics during surgery, while at the same time permitting the performance of an almost nontouch extirpation. Moreover, the hemostatic effect enables the surgeon to distinguish accurately between pathologic and normal tissue. Hence, the laser's application in cancer surgery is obvious. Those of us who have this modality at our disposal use it routinely for the excision of accessible malignant disease, in spite of the fact that the clinical follow-up is still too short to be able to reach definite conclusions regarding its applicability. Considerable experimental evidence exists, however, to indicate that the hypothesis upon which the use of lasers in cancer surgery is based is well founded.

Many surgeons have reported their experience in dealing with the surgical removal of cancer of various kinds and in various anatomic sites using the CO_2 laser. We have performed well over 200 wide excisions of malignant melanomas, with primary skin grafting, and have had no reason to regret having introduced this modality as a routine procedure in our department.

3.6 Operations Performed Through Highly Infected Tissue

The excisions of burns, synergistic gangrene, and decubitus ulcers are examples par excellence of the application of the CO_2 laser in this connection. Work in this field has been well documented and our experience confirms the advantage reported by others. Satisfactory results are also reported in the treatment of osteomyelitis while the perineal part of abdominoperineal resection when performed with the CO_2 laser leaves the patient with minimal postoperative discomfort and discharge.

3.7 Operations Performed on Organs Requiring Simultaneous Monitoring

The laser is especially useful in surgery on patients in whom monitoring is necessary. The fact that the heart and the brain can be monitored during laser surgery without interfering with the monitor has been shown to be of great advantage. In our hospital, the laser is used instead of electric cautery in surgery performed on patients with pacemakers in order to avoid possible cardiac arrest.

3.8 Cavitational Surgery

The use of the CO_2 laser with microscope attachments and its advantages over other modalities is well established. In our hospital, surgery of the uterine cervix has been performed with satisfactory results using the CO_2 laser combined with a colposcope. Moreover, rectal surgery has also been shown to be practical using this modality. Urologic surgery employing a cystoscope is, as yet, in the experimental stage, but there is every reason to believe that its clinical application will be established in the near future. Microneurosurgery is being successfully conducted and there is no doubt that this technique will be universally accepted in the near future. Tubal surgery performed through a laparoscope has become a routine procedure in our hospital. The same applies to laryngeal and bronchial surgery using the microscope attachment or a bronchoscope.

3.9 Specific Tissues Best Incised by Means of the CO_2 Laser

Excellent examples of this application are incisions performed through the sclera and spinal meninges. In the former, the laser prevents an increase in intraocular pressure and hemorrhage into the vitreous; in the latter, manipulation of the spinal cord with resultant damage to the nerve roots is avoided.

When one considers that extirpative surgery performed with the CO_2 laser on a clinical basis commenced only 10 years ago one cannot help feeling that its application in surgery will become more and more universal as more surgeons introduce this modality into their armamentarium. The advantages of the CO_2 laser in private practice will be stressed in the chapter on Dermatological Surgery.

3.10 Precautions

Since the beam can theoretically be reflected from a shiny instrument back toward the eyes of those in the immediate vicinity of the operating field we recommend that those in the proximity of the CO_2 laser wear spectacles (optical lenses are not necessary since the infrared beam cannot penetrate glass). This reflection in our experience is purely theoretical since by the time the reflected beam reaches the eyes it is completely diverged and relatively harmless. Nevertheless, we recommend that anodized instruments be used in the operative field to avoid reflection.

To protect adjacent tissue from inadvertant damage, this tissue can be shielded with wet gauze or a metal instrument can be interposed between the tissue being incised and that to be protected. It is essential, therefore, to cover intratracheal tubes with aluminium tape during laryngeal surgery in order to prevent ignition. Inflammable anesthetic gases must never be used during laser surgery. This also applies to materials used for preparing the skin.

3.11 Operations in Which the Anticipated Blood Loss is Significant

3.11.1 Breast Surgery

Mastectomies and mammaplashes have been performed with the object of minimizing blood loss. The results have in our opinion justified this approach both from the point of view of the reduced blood loss and for cosmetic reasons.

Figure 4 a–d

(a) *Intraoperative view of a mastectomy. Note the wound after excisional biopsy performed with the CO_2 laser so that a frozen section could be obtained*

(b) *Intraoperative condition after modified radical mastectomy. Note dry healthy operative field. No ligatures were required. The spots of carbonized tissue indicate larger blood vessels, which were sealed by clamping and vaporizing the tissue with a defocused beam between the jaws of the clamp. In these operations we have found the insertion of drains to be unnecessary*

(c) *Intraoperative view of mastectomy showing the axillary dissection*

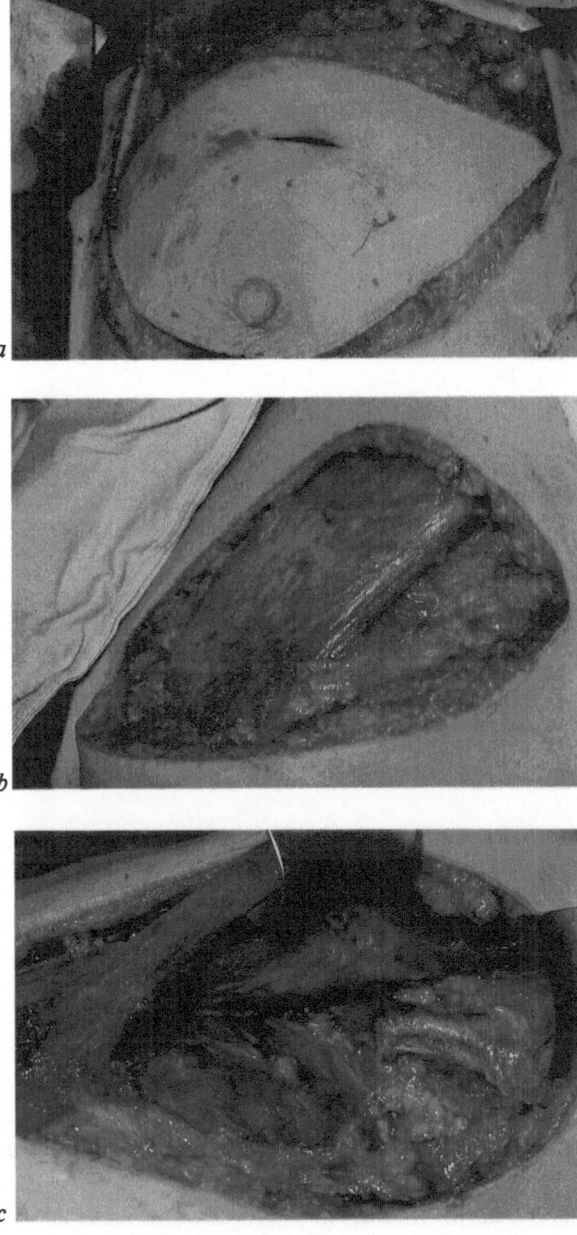

Figure 4
(d) Postoperative view after 6 months

Figure 5 a–d

(a) Bilateral breast cancer. In this case, a comparison between laser and conventional surgery was made. It was found that not only was the blood loss significantly less on the laser side but the postoperative course was greatly superior with regard to pain and mobility. In this case, the laser side took approximately 15 min longer to perform; this, however, is not always so

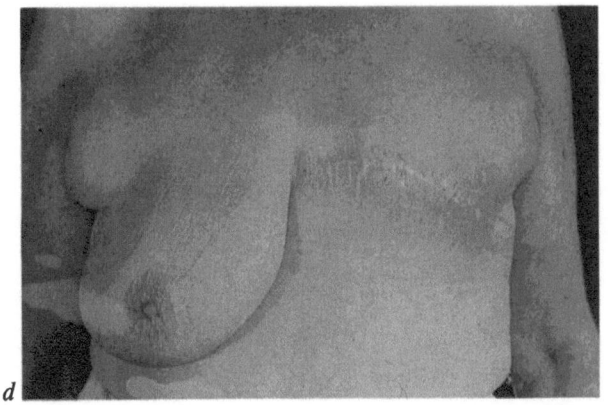

4d

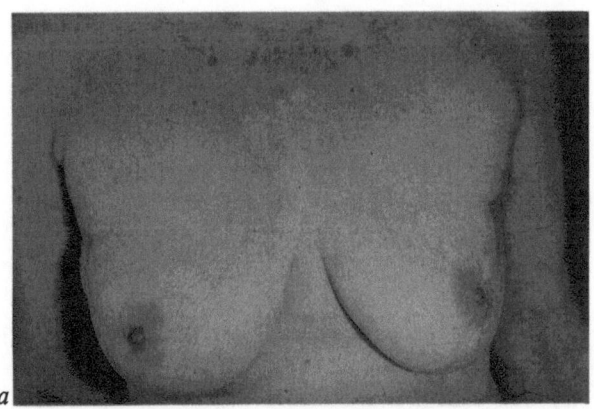

5a

Figure 5

(b) *The immediate postoperative view*
(c) *The condition 10 days postoperation, after removal of sutures. Note no difference in wound healing*
(d) *The condition 6 months later*

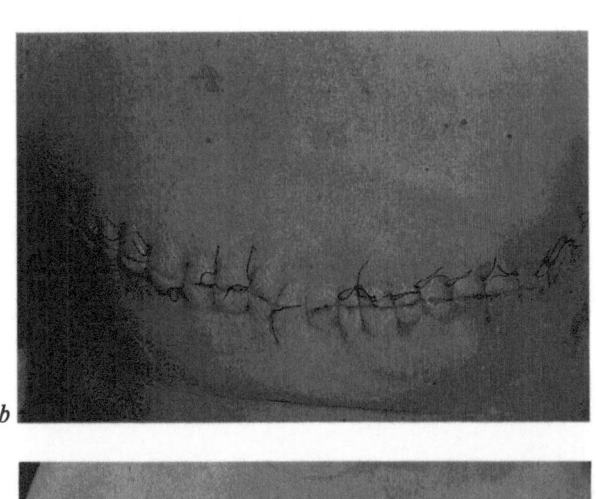

b

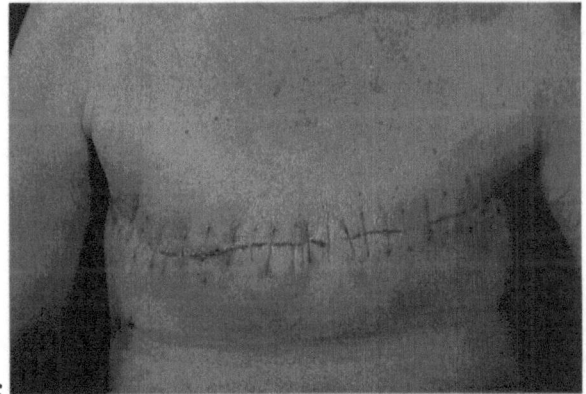

c

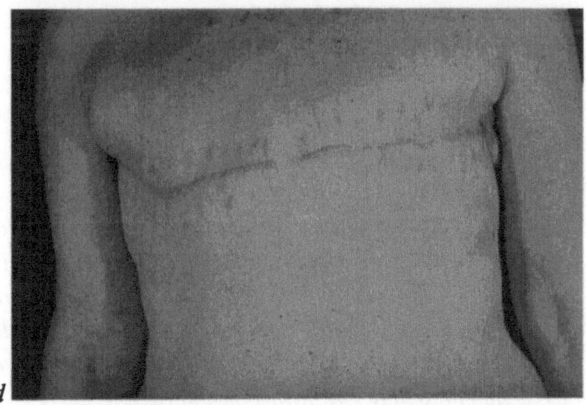

d

3.11.2 Palliative Surgery for Incurable Malignant Diseases

Surgery in cases with cutaneous metastases and large incurable cutaneous tumors may result in massive bleeding, infection, sloughing of the wound, sepsis, and pulmonary complications because the tumors usually ulcerate or fungate as a result of infection. The fact that the CO_2 laser seals blood vessels, lymphatics, and nerve endings and that the highly sterilizing effect of the laser beam results in the ability to perform operations upon highly infected tissue with decreasing morbidity justifies this approach. Although surgery in these cases can only be palliative, it benefits the patients and their families and can provide a tolerable existence even though survival may not be prolonged.

Figure 6 a–c

(a) *A case with incurable breast cancer, where although the tumor had been present for 4 years the patient refused surgery until massive bleeding, malodor, and severe discomfort drove her to seek medical care. A wide excision of the tumor with axillary lymph gland dissection was performed*

(b) *The view 24 h after surgery. Note: erythema in the region of the operation is absent. This is due to the fact that the skin incision was performed with high energy density and rapid scanning. Although a drain was inserted in this case it was removed the following day because it was found to be unnecessary*

(c) *The condition 1 month postoperation*

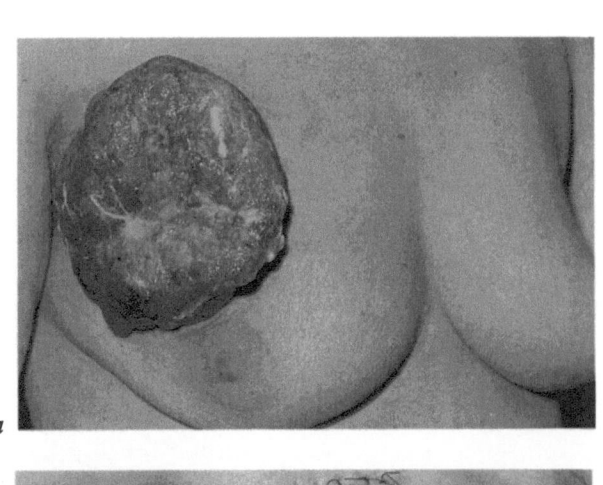

a

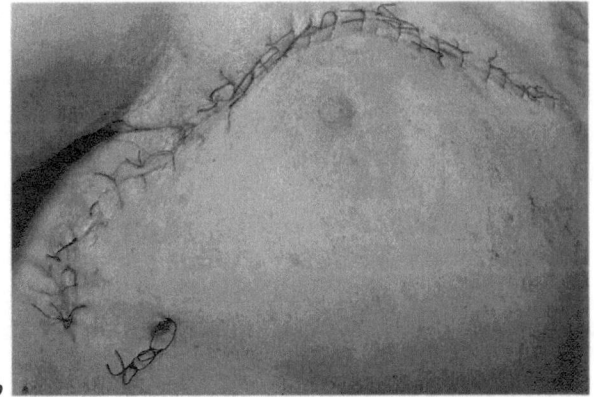

b

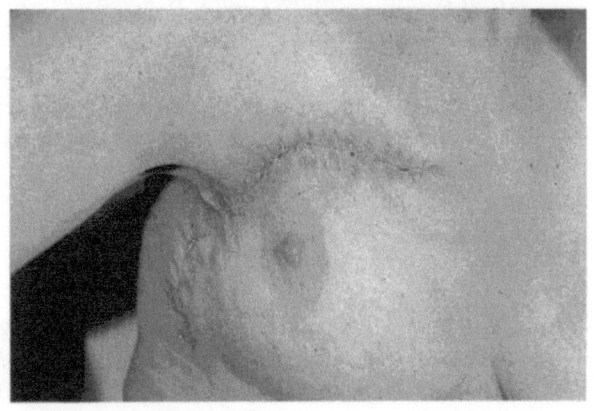

c

Figure 7 a–c

(a) *Another case of incurable breast cancer widely excised and grafted with a split-thickness skin graft. According to our clinical experience, there is a slight delay in the body's acceptance of split-thickness skin grafts when operations are performed with the laser. To overcome this, delayed primary grafting is used. Wet dressings (sterile saline) are put on the wound surface for 12 h, and then the skin grafting is carried out. Using this method, in most cases there was a highly satisfactory acceptance of the grafts and in only a few cases was the acceptance around the edges less complete*

(b) *One week after the operation. The graft was placed on the wound and left exposed. No dressing or tie-over pack is necessary, nor is it necessary to suture the graft into position since the CO_2 laser appears to have several advantages in preparing a site for split-thickness skin grafts – good hemostasis and sealing of the lymphatics reduces the separation of the graft through accumulation of blood or other fluids and the minimal tissue damage produced by the laser beam reduces the tissue necrosis and thus the risk of infection*

(c) *The condition 1 month post operative*

Figure 8 a and b

(a) *A patient with a recurrent breast cancer 4 years after radical mastectomy*

(b) *Radical excision and split-thickness skin grafting was performed. The graft was placed on the periosteum of the ribs. The condition 2 years after the operation*

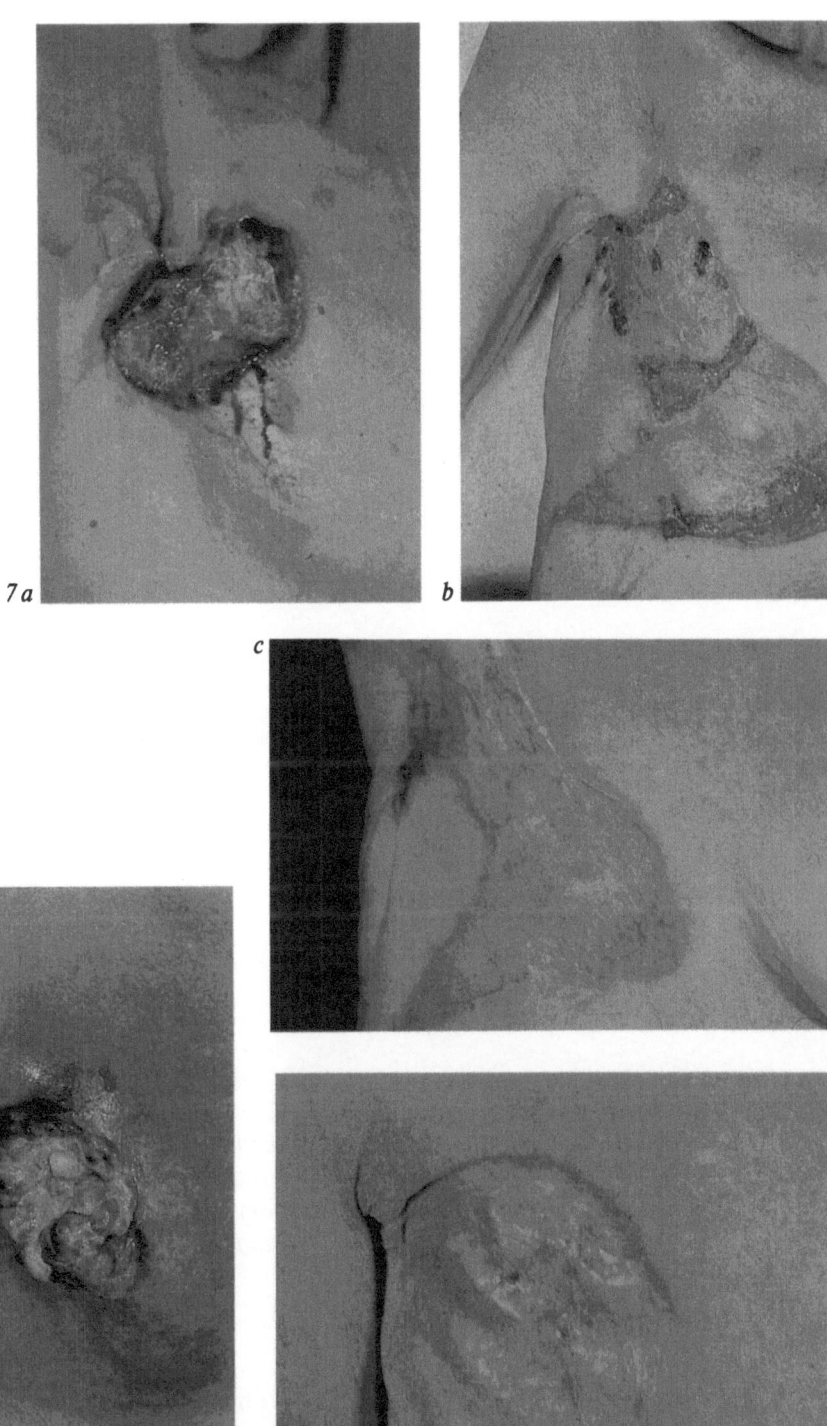

Figure 9 a and b

(a) *Recurrent adenocarcinoma of the rectum (postabdominoperineal resection) in the anal region*

(b) *The view 3 months after excision and primary suture. It is worthy of mention that small recurrences in the anorectal region can be vaporized if excision is contraindicated*

Figure 10 a and b

(a) *A case of locally spreading seminoma with no response to radio- and chemotherapy. Excision of the tumoral mass with primary repair of the wound was performed*

(b) *The condition one week after operation*

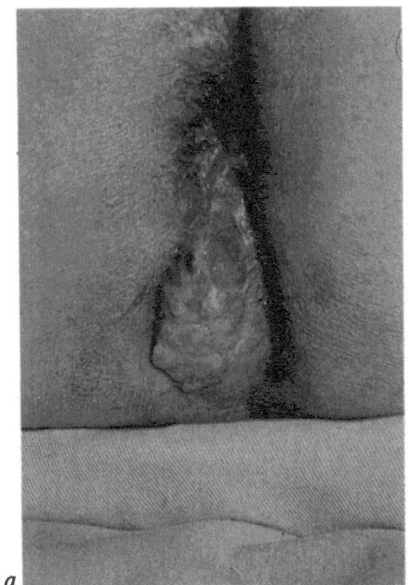
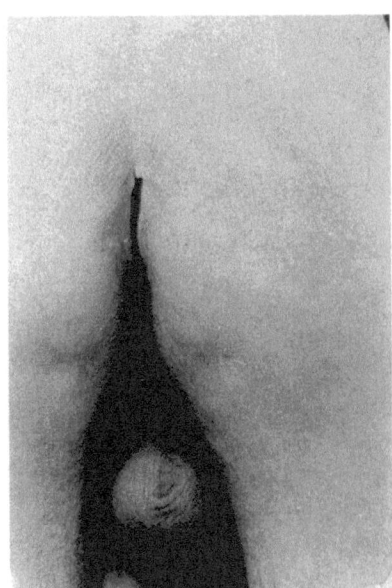

9 *a*

b

10 *a*

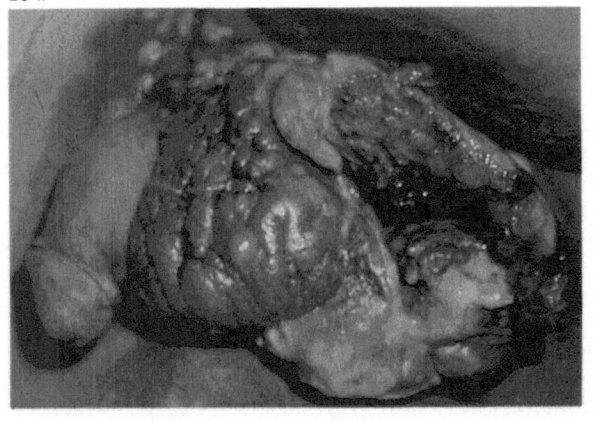

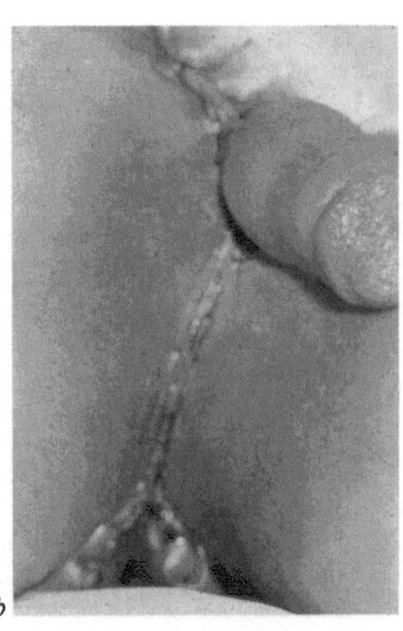

b

Figure 11a–c

(a) *Recurrent fibroliposarcoma of the thigh*
(b) *The condition immediately after wide excision. Note how clearly the anatomy can be identified due to lack of bleeding*
(c) *Two days postoperation, mesh split-thickness skin graft covers the large defect*

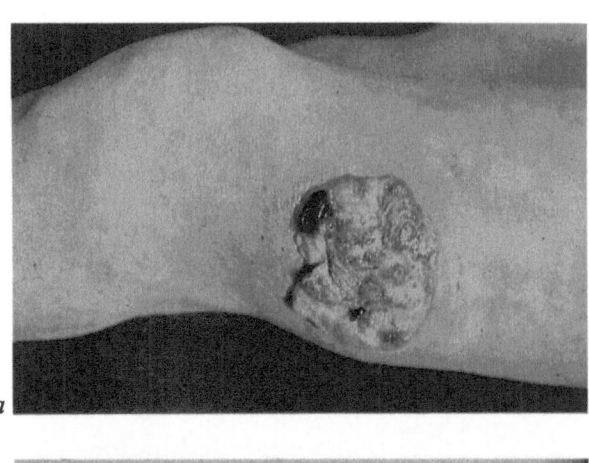

a

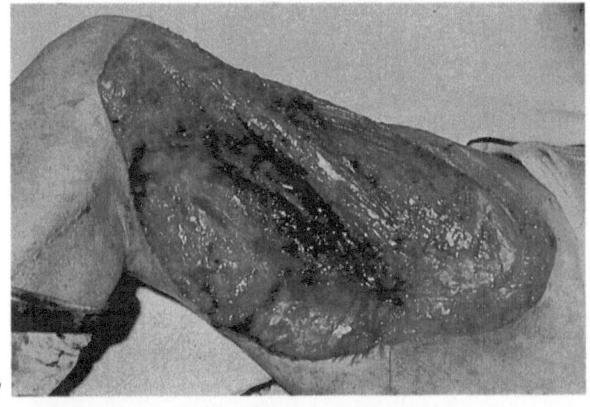

b

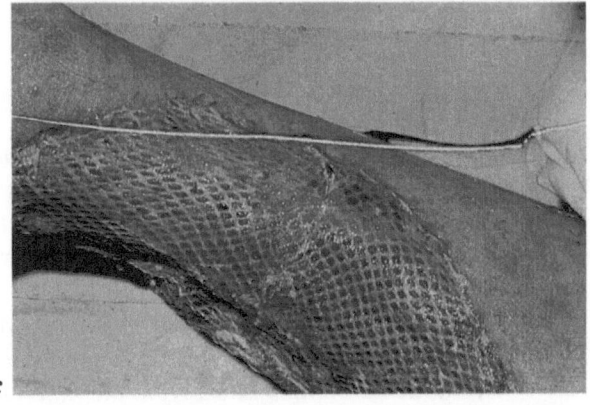

c

3.11.3 Neonatal Surgery

One of the established advantages of CO_2 laser in surgery is its hemostatic effect, which minimizes intraoperative blood loss. Nowhere is this more important than in surgery performed on the newborn – so much so that many of us firmly believe that this modality should be used in all surgical procedures performed on the neonate.

Figure 12 a–d

(a) *Preoperative view of a baby with lymphangioma of the left chest wall, shoulder, and upper third of the arm, infiltrating deep to the scapula. The baby was in respiratory distress due to the pressure of the tumoral mass on the chest wall*

(b) *The condition at completion of extirpation. Note the brachial plexus, which was dissected out with the laser*

(c) *The condition at completion of surgery. The two "dog-ears" were left intentionally in order not to compromise the blood supply to the skin flaps*

(d) *The condition 2 months later. Note the functional result in the affected arm*

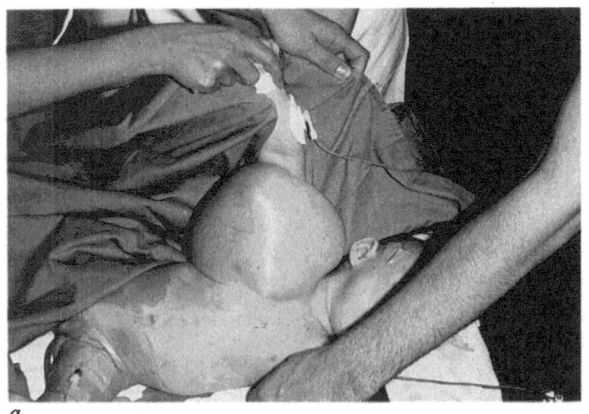

a

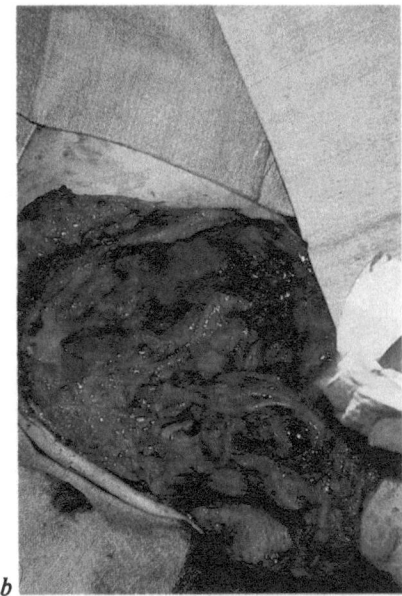

b

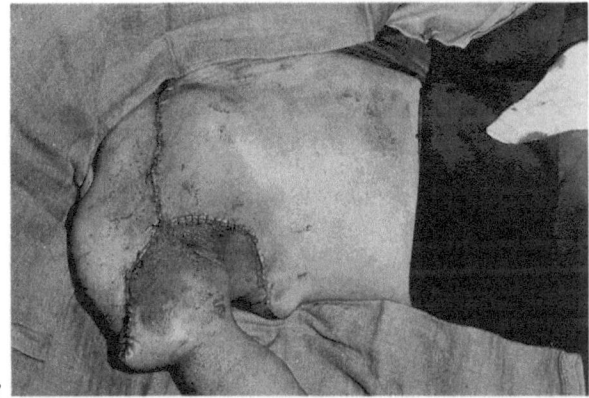

c

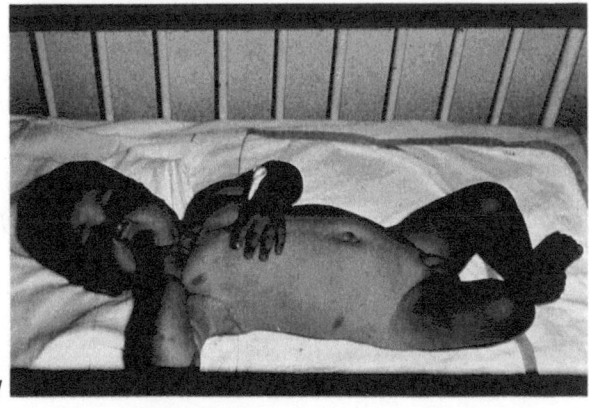

d

Figure 13a–d

(a) *One of a pair of identical twins was born with a trouser-shaped giant nevus asso-ciated with large neurofibromatous tumors of the buttock, which obstructed the anus and made nursing almost impossible*

(b) *A closeup view of the neurofibromatous tumors. The tumors were excised with the CO_2 laser with negligible blood loss; the postoperative course was uneventful*

(c) *The condition after 4 months*

(d) *The same boy 2 years later*

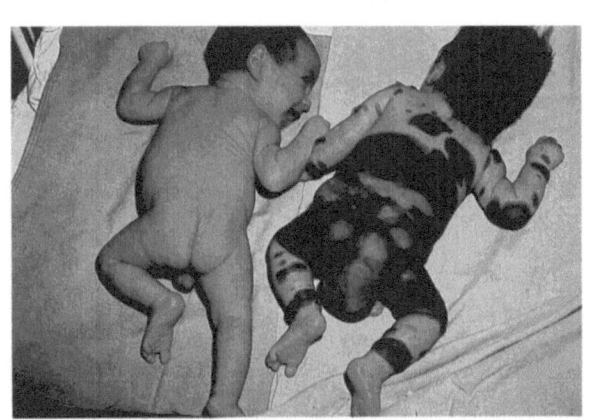

a

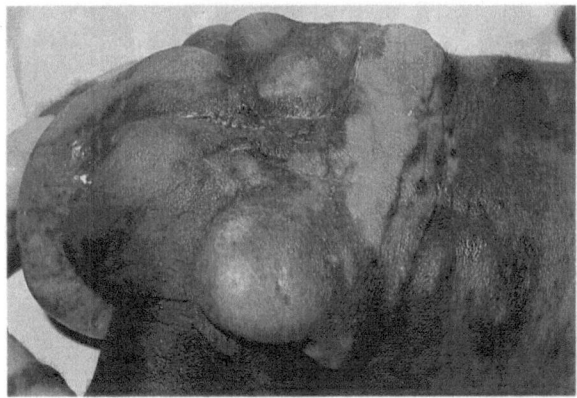

b

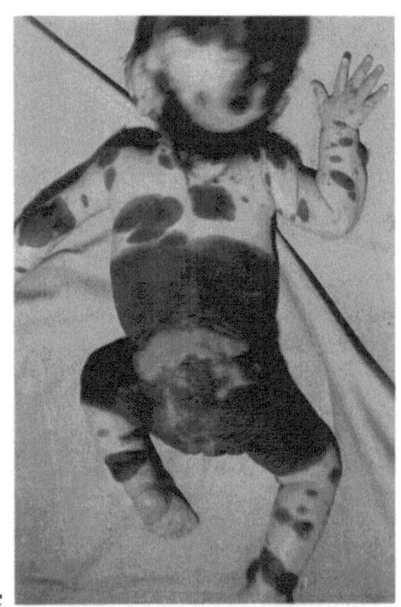

c

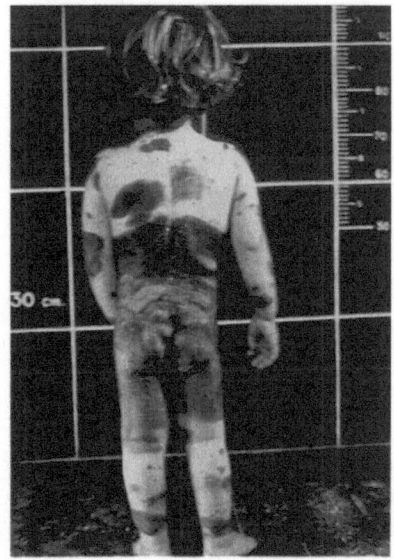

d

Figure 14a–c

(a) A child born with a sacrococcygeal teratoma. The tumor was excised in the imme-
     diate postnatal period
(b) The condition 2 weeks post operative
(c) The condition 4 years later

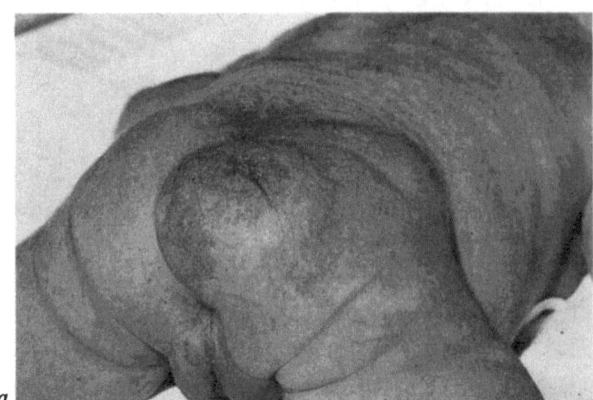

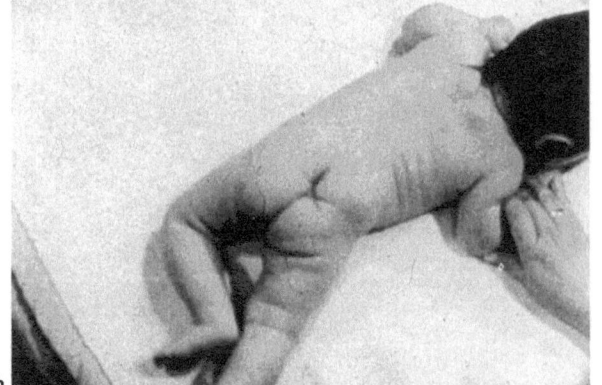

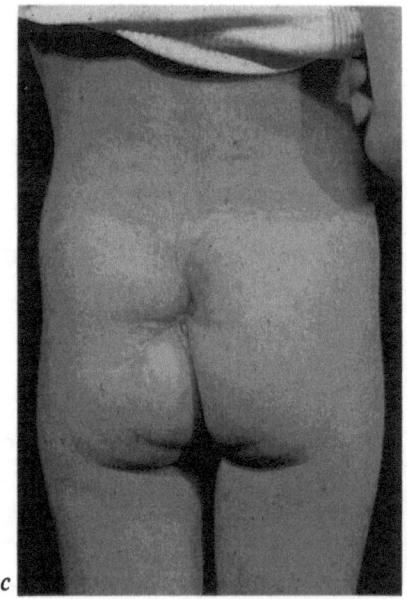

3.12 Surgery Performed on Highly Vascular Areas of the Body

3.12.1 Scalp Surgery

The advantages of using the CO_2 laser on the scalp are related to the hemostatic and sterilizing effects of the beam. Surgery on the scalp can, therefore, be performed with minimal shaving and preparation. Since suturing split-thickness skin grafts into position is unnecessary, the operation as well as the postoperative dressing and care are simplified.

Figure 15a and b

(a) Squamous cell carcinoma of the scalp
(b) The condition after wide excision and split-thickness skin grafting. The graft is left exposed

Figure 16a–c

(a) A large basal cell carcinoma of the scalp
(b) Twenty four hours after the operation, a split-thickness skin graft is placed on the surgical bed and left exposed
(c) The condition 1 year after the operation

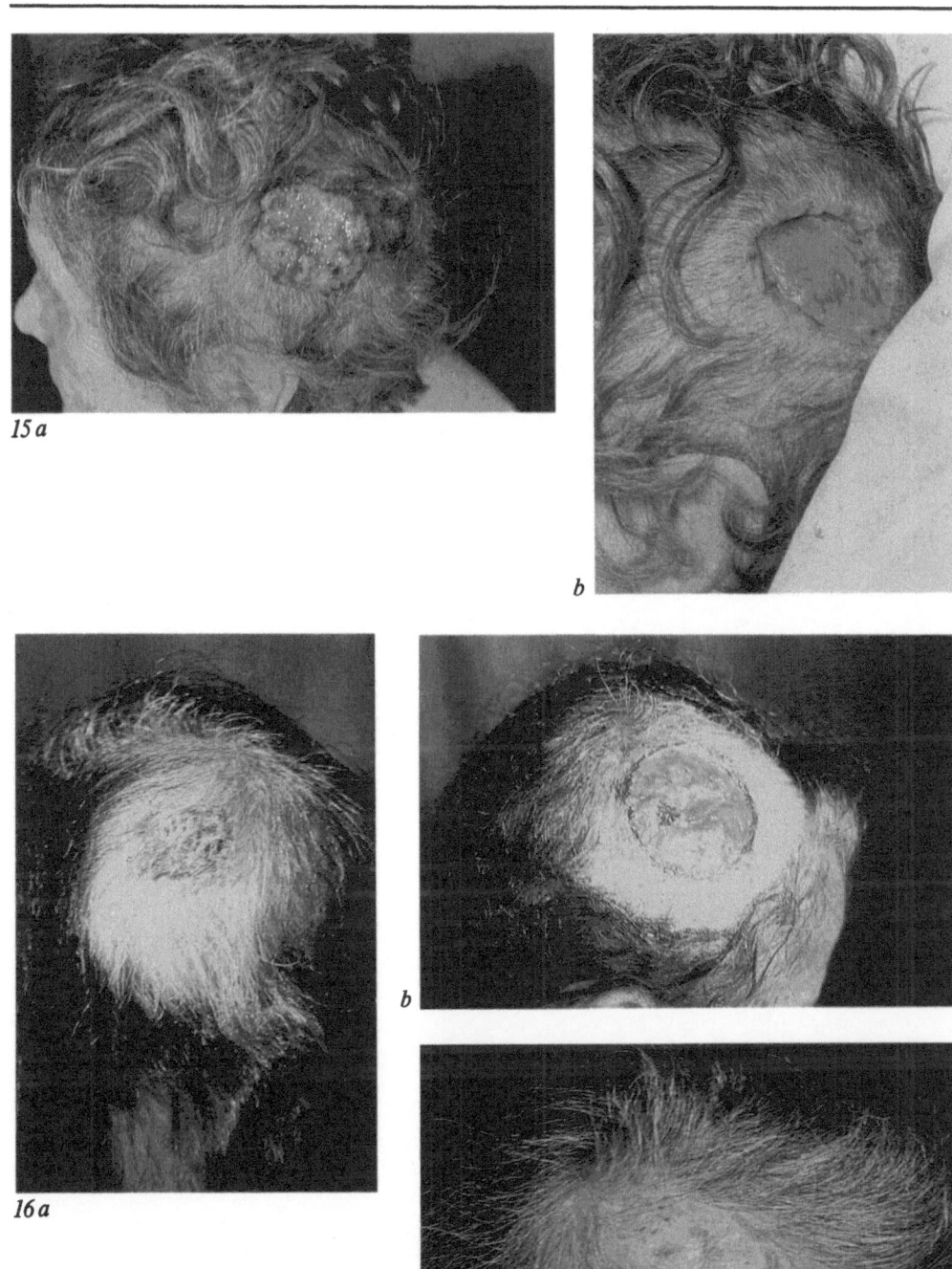

15 a

b

16 a

b

c

Figure 17a and b

(a) Squamous cell carcinoma of the forehead. Excision and primary split-thickness skin grafting were performed
(b) The condition after 1 year

Figure 18a and b

(a) Recurrent fibrosarcoma of the occipital region. Excision and repair by rotation flap were performed
(b) The condition 1 month after the operation

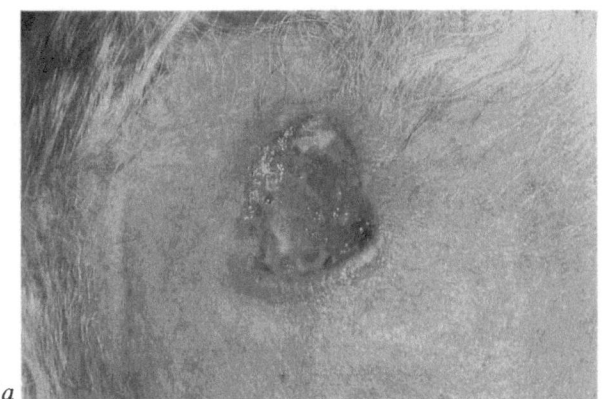

17a

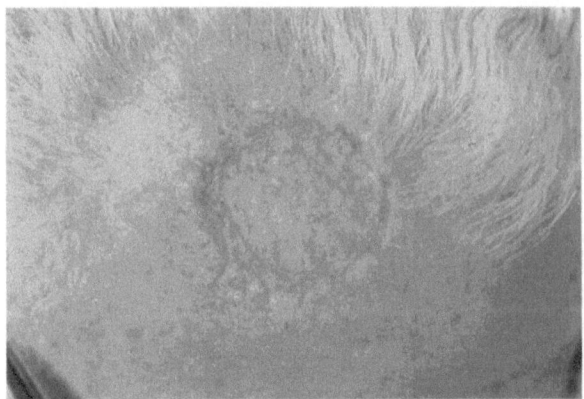

b

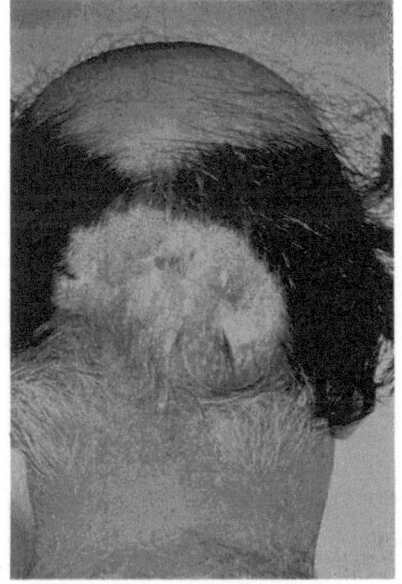

18a

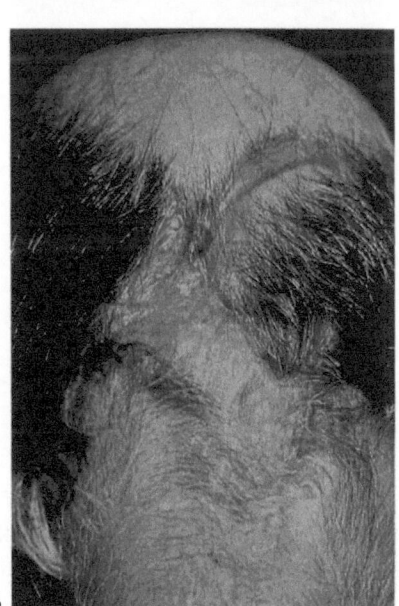

b

3.12.2 Tongue Surgery

Apart from the hemostatic effect of the CO_2 laser, which reduces the intraoperative bleeding, the absence of postoperative oozing enables a graft to be placed on the defect. The lack of postoperative pain and discomfort make this type of surgery advantageous. We have also found that after laser surgery in the oral cavity and oropharynx, there is minimal postoperative edema so that in the vast majority of cases a tracheostomy can be avoided.

Figure 19a–c

(a) Leukoplakia along the left side of the tongue
(b) Immediate postoperative view. Excision was performed and a split-thickness skin graft was sutured to cover the defect
(c) Fourteen days after the operation showing a complete skin cover of the defect

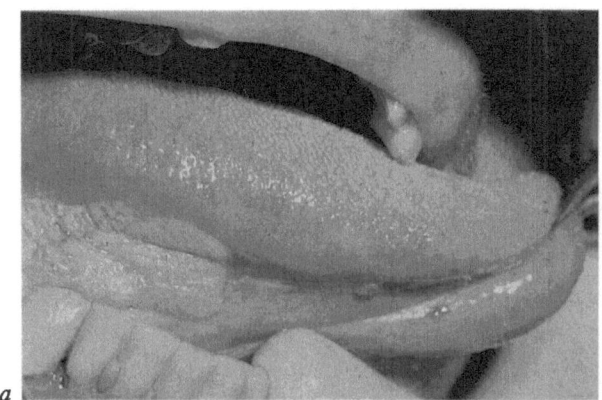

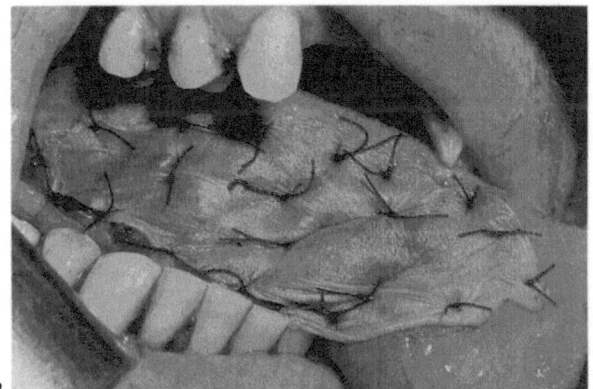

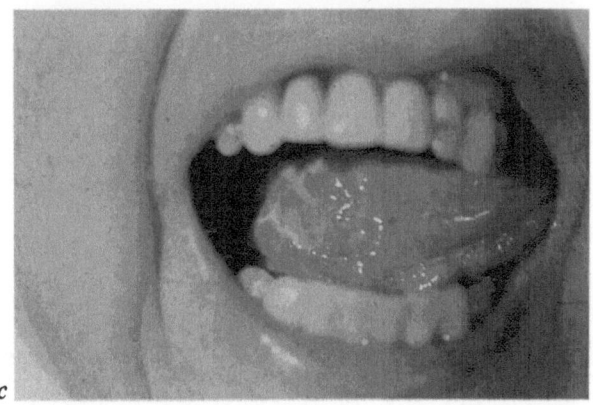

Figure 20a and b

(a) *A 14-year-old girl with macroglossia. Her mouth was continuously open with the tongue protruding*
(b) *Following completion of excision there was no blood oozing from the wound. No ligatures were necessary. The wound was sutured directly. Complete healing ensured*

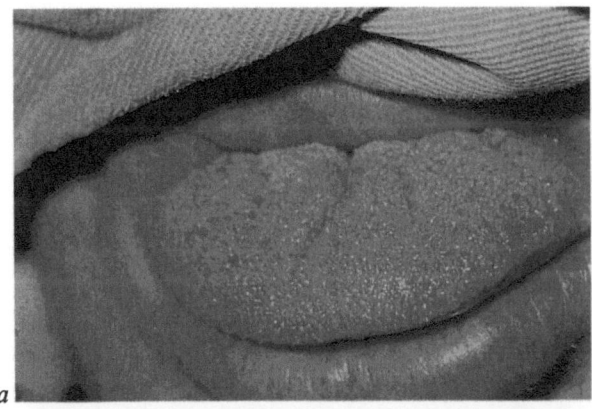

a

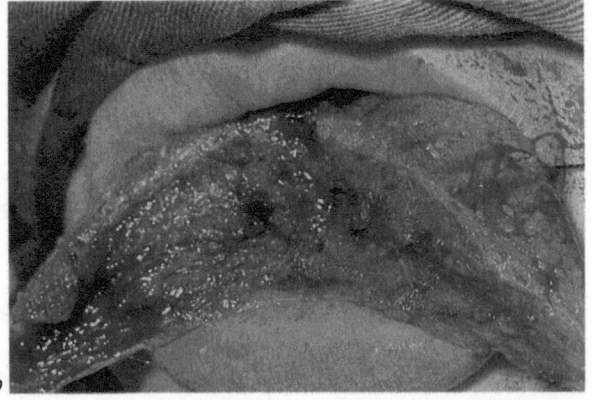

b

Figure 21a–d

(a) *A patient with Down's syndrome. Notice the difficulty with which the mouth is kept closed because of the macroglossia*
(b) *The macroglossia of the same patient. The marking shows the extent of excision. Excision and primary closure of the wound were performed*
(c) *The condition 2 months after surgery*
(d) *The patient after the operation, there is no difficulty in closing the mouth*

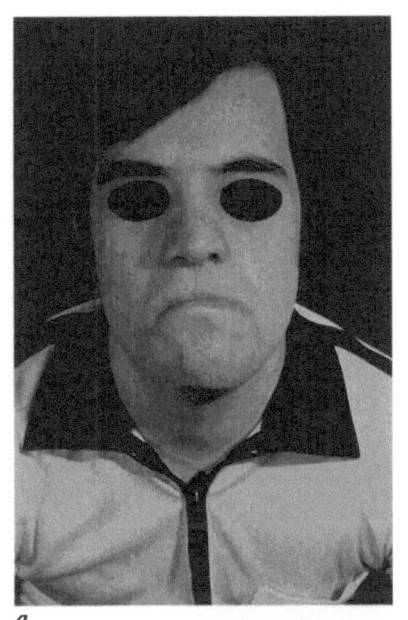

a

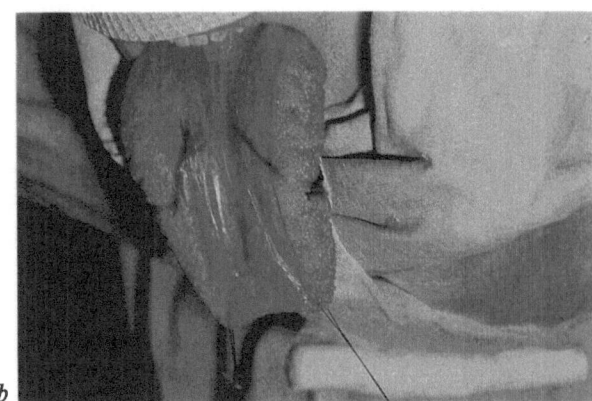

b

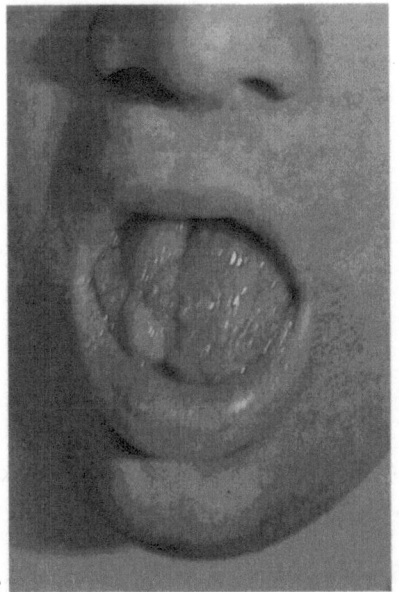

c

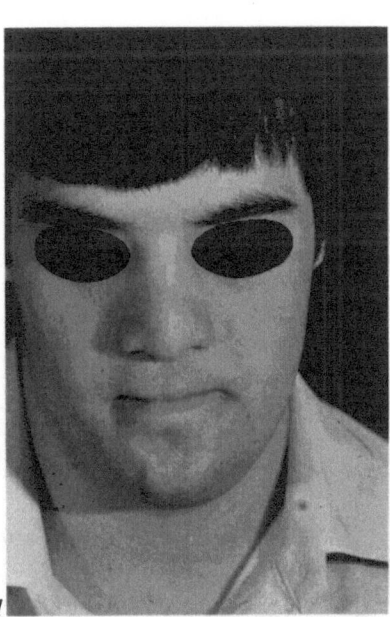

d

Figure 22a–c

(a) *Squamous cell carcinoma of the lateral aspect of the tongue. The sutures indicate the site of previous biopsy taken without the laser beam. Wide excision with primary suturing was performed*
(b) *The resected specimen*
(c) *The condition 3 months after the operation indicating complete healing*

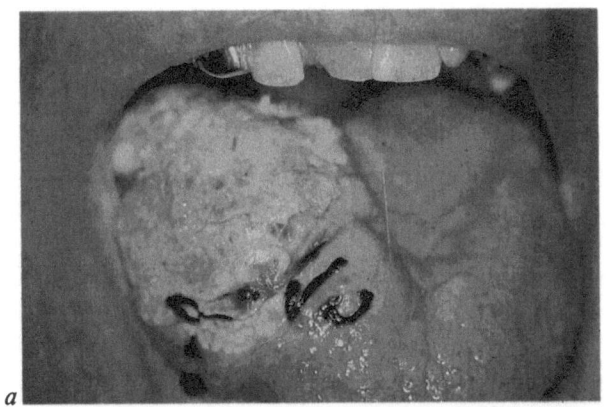

a

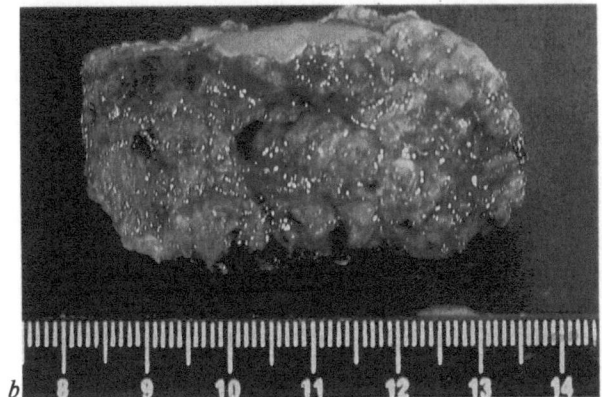

b

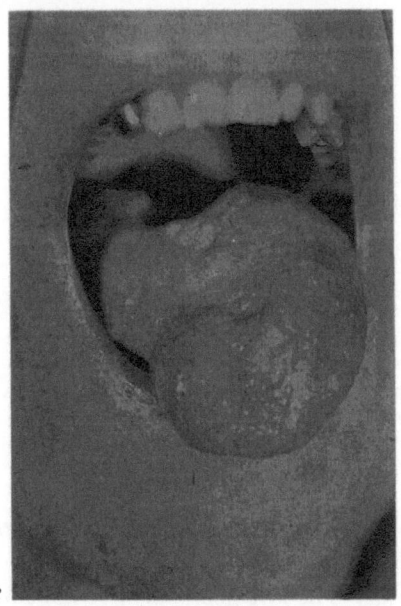

c

3.12.3 Extirpation of Highly Vascular Tumors

Strawberry nevi and cavernous hemangiomas are among the most common tumors in childhood and most of them involve the head and neck regions. Strawberry nevi usually involute spontaneously by the age of 5 years. Those associated with an underlying cavernous element only dissappear partially and the subcutaneous element requires extirpation. Cavernous hemangiomas, however, do not tend to involute and require an active approach of which surgical extirpation is the treatment of choice. Although in most cases early therapeutic treatment is contraindicated, the need for active treatment should be considered in cases of thrombocytopenia and disseminated intravascular coagulopathy or when ulceration and hemorrhage occur. Other indications are when a tumor interferes with normal activities, when permanent functional impairment could ensure, especially in the face, and in lesions with atypical growth patterns. The advantages of the CO_2 laser beam have led us to use this modality in the surgical treatment of all cases with hemangiomas.

Figure 23a and b

(a) Strawberry hemangioma of the upper lip. The lesion was excised with negligible blood loss
(b) The late postoperative appearance

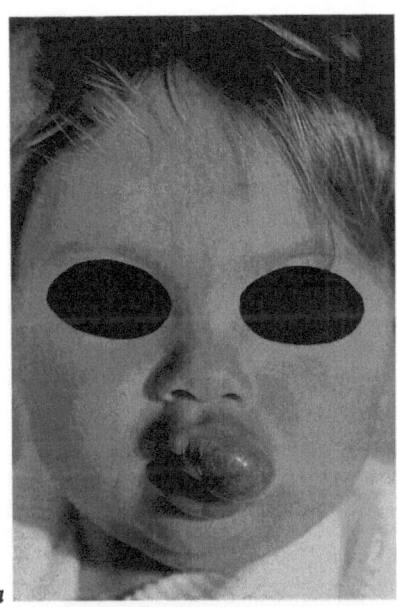

a

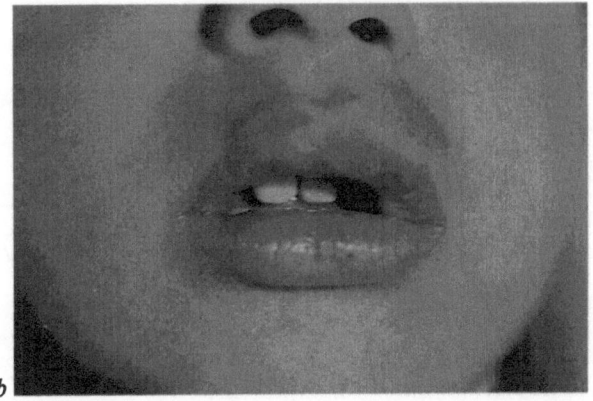

b

Figure 24a–c

(a) A similar case to the one in Figs. 23a and b. Surgery was delayed until the capillary element had regressed

(b) The intraoperative view after extirpation of the cavernous element showing no bleeding

(c) The condition 3 months later

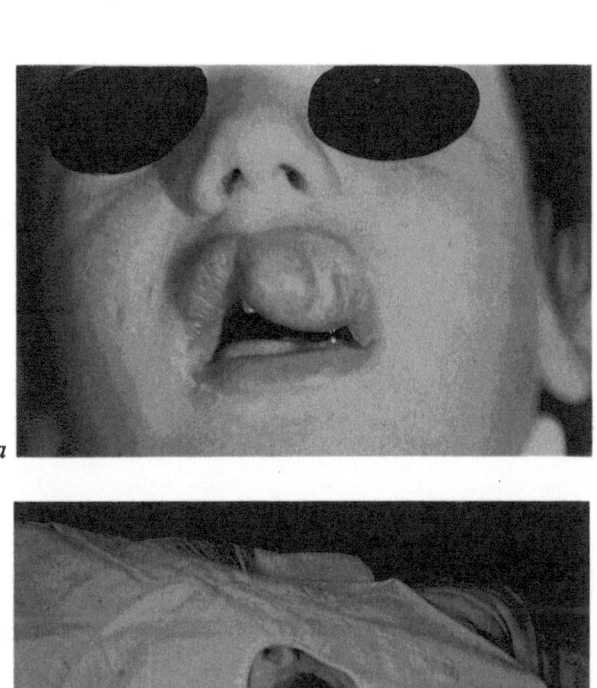

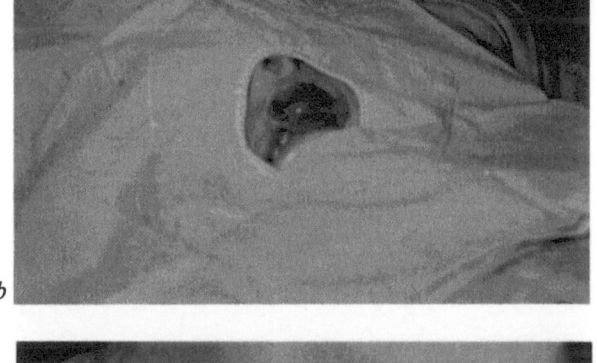

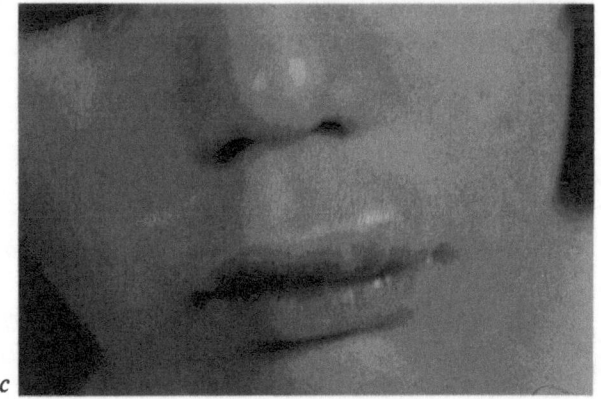

Figure 25 a–g

(a) A female baby with a strawberry nevus of the glabellar region

(b) The patient in Fig. (a) at the age of 1 year; in spite of inactivity, the cavernous element remained. The hemangioma in this case had infiltrated between the nasal bones

(c) At the age of 4 years the operation was performed. Access was gained without outward fracturing of the nasal bones and the part of the cavernous element between the nasal bones was vaporized

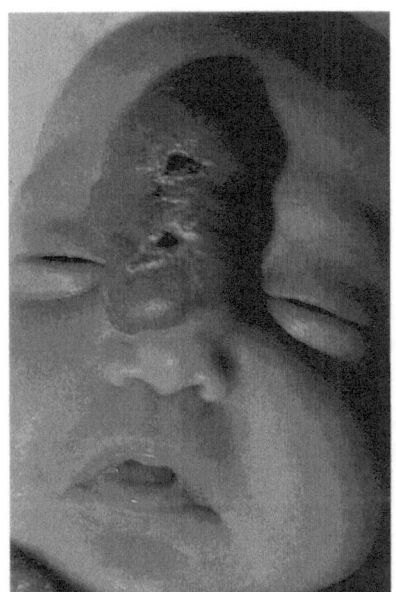

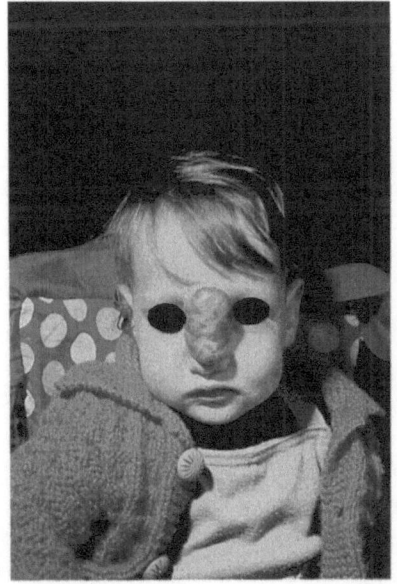

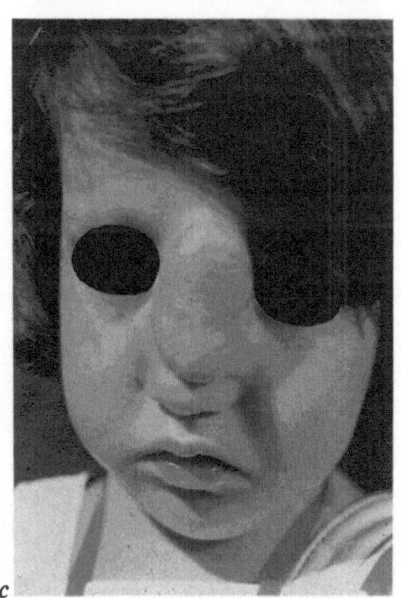

Figure 25

(d) The condition 1 month after operation
(e) The condition 1 year after operation
(f) The condition 4 years after operation
(g) The lateral view showing normal growth of the nose

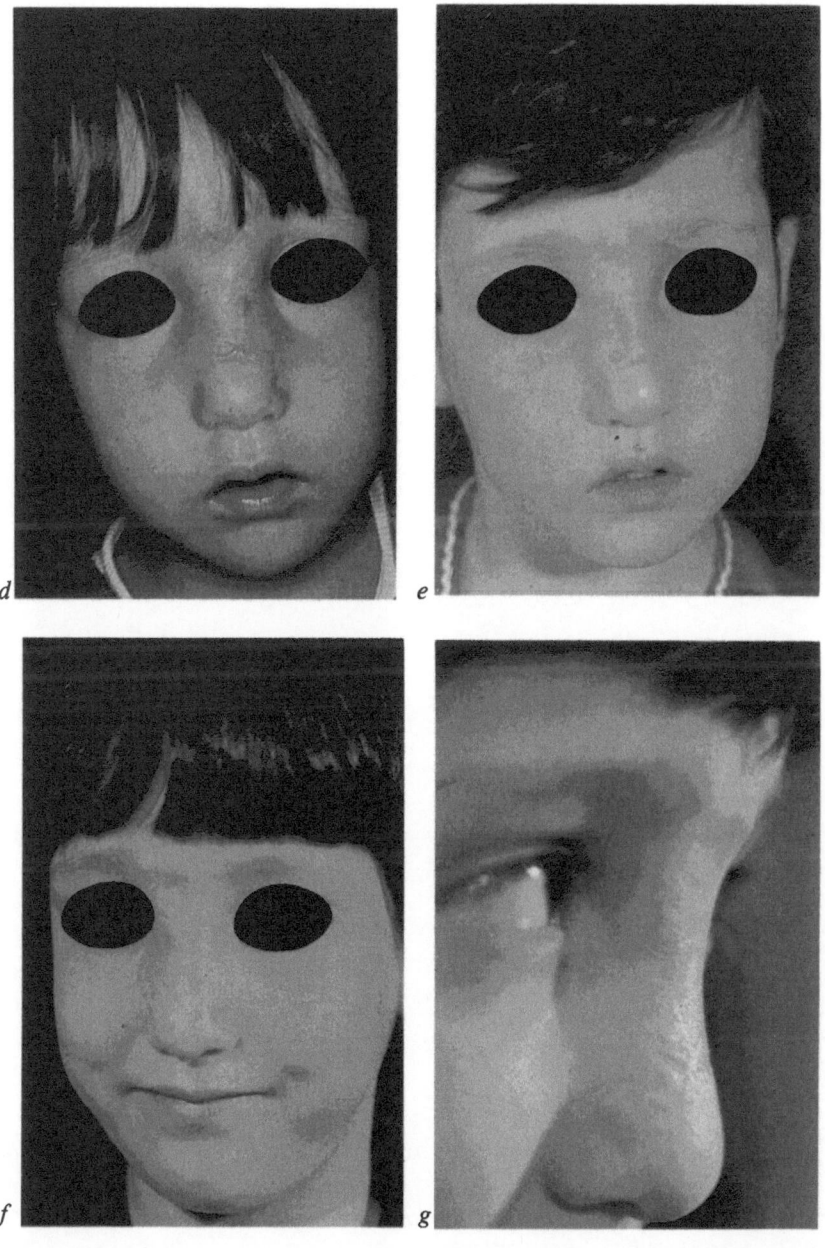

Figure 26 a and b

(a) A 2-year-old boy with a strawberry hemangioma on the lateral aspect of the nose

(b) The condition 6 months after laser extirpation

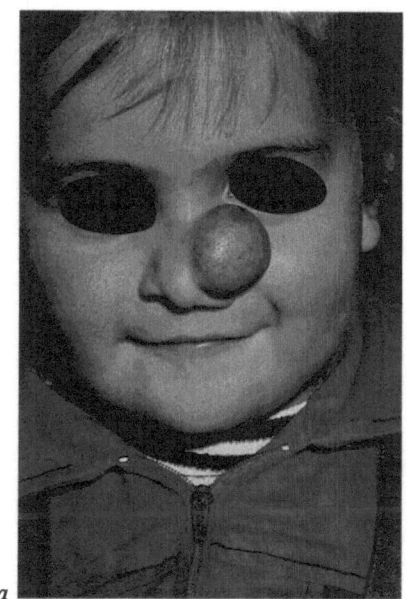

a

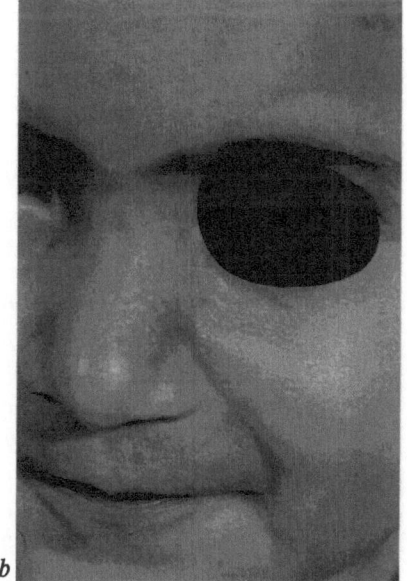

b

Figure 27 a–d

(a), (b) *A female baby with hemangioma involving the whole of the left side of the face extending from the postauricular to the submental areas. The extirpation of the hemangioma was performed through a facelift incision. It should be mentioned in this case that the lack of intraoperative hemorrhage enabled us to identify the anatomy very accurately. However, the facial nerve was seen to enter the tumor and had to be partially resected. The eventual cosmetic and functional results, though, were acceptable*

(c), (d) *Anterior and lateral views of the condition 3 months after the operation. Note the minimal scarring of the face-lift incision*

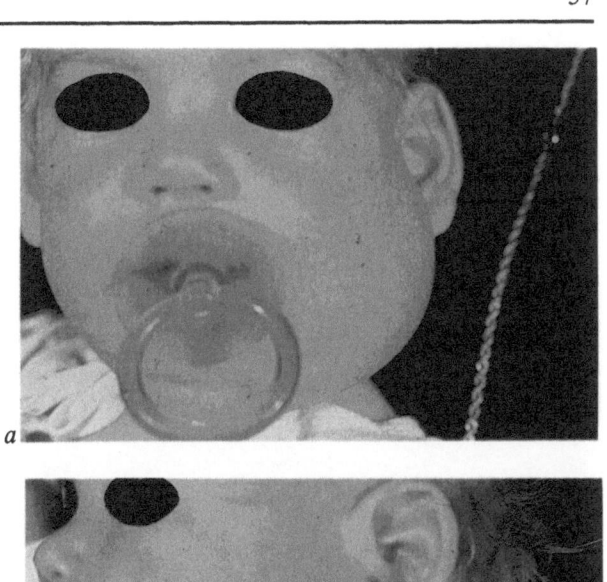

a

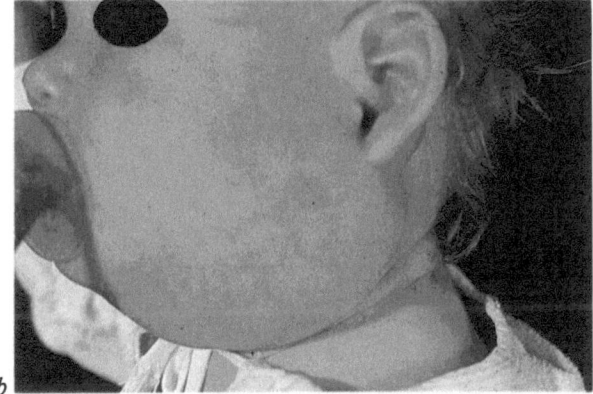

b

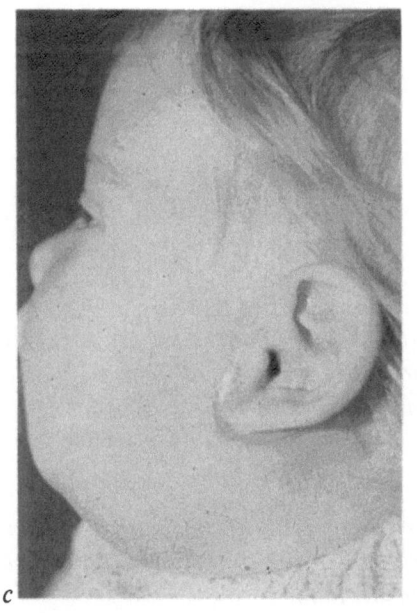

c

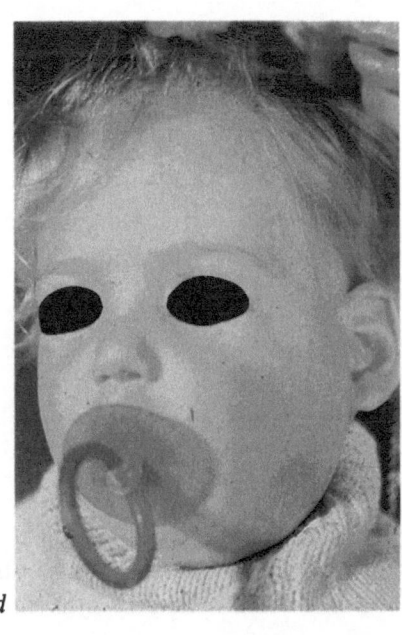

d

Figure 28a and b

(a) *An auricular cavernous hemangioma; several attempts at excision were abandoned because of copius bleeding, which was extremely difficult to control*
(b) *The condition 2 months after laser excision*

Figure 29a and b

(a) *Cavernous hemangioma of the entire auricle. In this case, excision was performed with split-thickness skin grafting for closure of the large skin defect*
(b) *The condition 1 year after the operation*

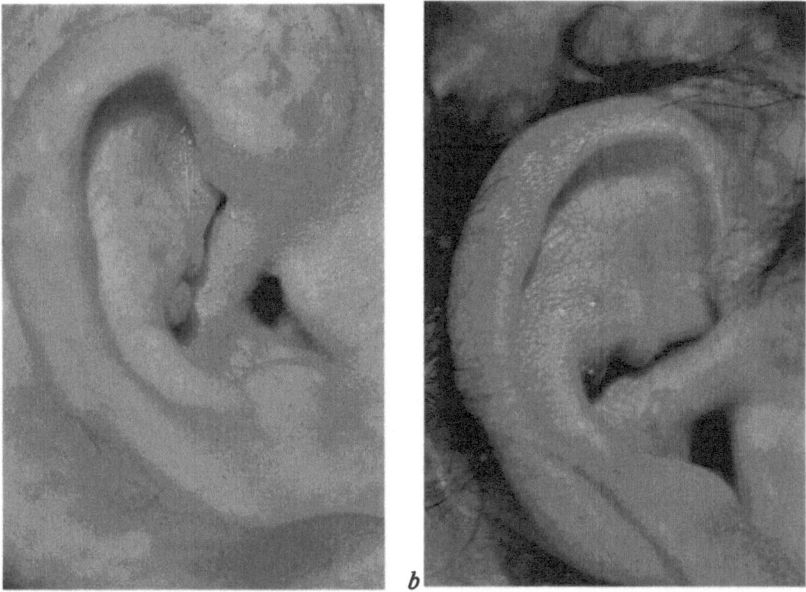

28 a   b

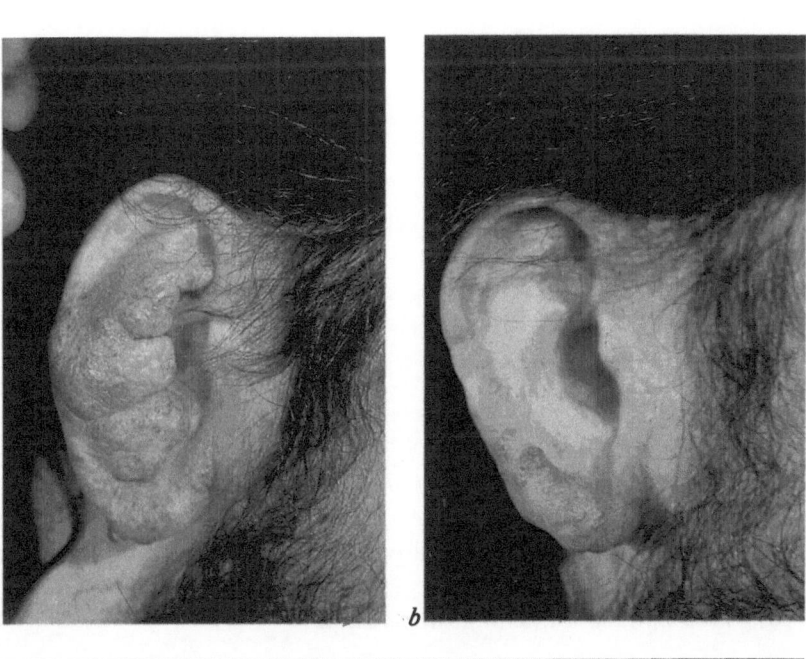

29 a   b

Figure 30a–c

(a) *A cavernous hemangioma of the vulva and perineum. In this case, surgery was delayed until adulthood for fear of excessive uncontrollable bleeding*

(b) *The immediate postoperative condition. The blood loss was measured and found to be approximately 50 cc*

(c) *The condition- 1 month postoperation*

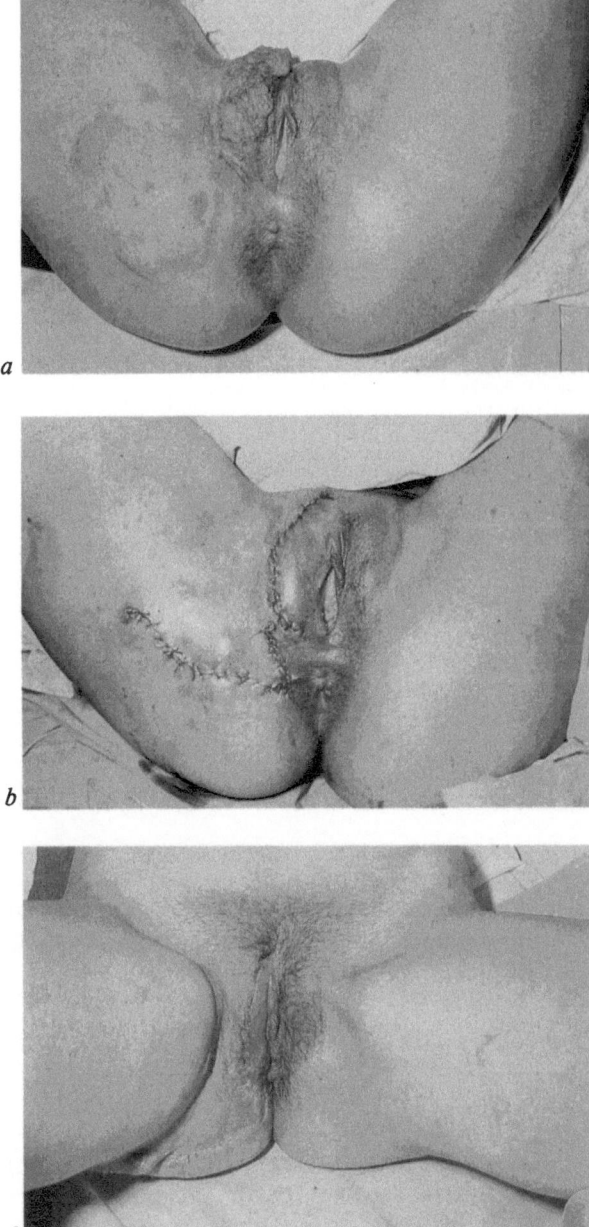

Figure 31a and b

(a) *Pedunculated cavernous hemangioma of the vulva in a 2-year-old girl. The pedicle was transected and the base was vaporized. No sutures were necessary*

(b) *The condition 5 days after the operation*

Figure 32a and b

(a) *Multiple hemangiomatous excrescences in a long-standing port-wine stain. Apart from the cosmetic deformity, excessive spontaneous hemorrhages led this patient to seek medical care. The large lesions were partially excised whereas the small lesions were vaporized by a defocused beam*

(b) *The postoperative condition after 3 months*

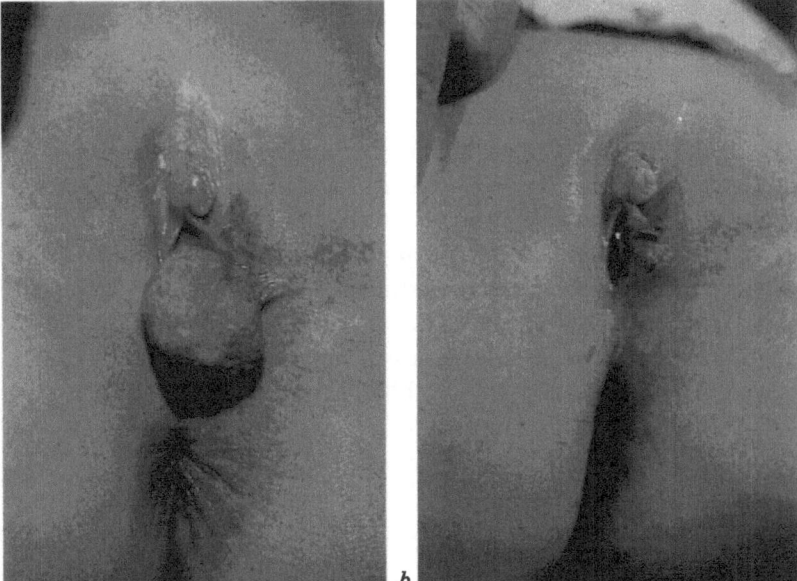

31 a   *b*

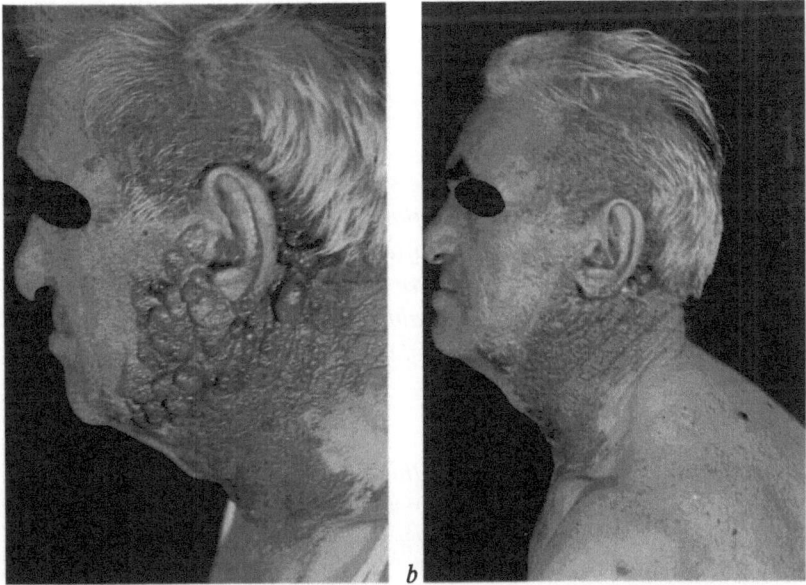

32 a   *b*

3.13 Surgery for Malignant Diseases

3.13.1 Malignant Melanoma

The accepted surgical management of malignant melanoma includes wide excision of the primary tumor and split-thickness skin grafting to cover the skin defect. When the regional lymph nodes are clinically palpable and suspected of harboring metastases, or when the depth of invasion of the primary lesion is level four or deeper according to Clark's classification, it is generally agreed that regional lymph gland dissection is indicated. Following lymph gland dissection, especially in the inguinal region, lymphorrhea, skin sloughing, and infection may occur. The ability of the CO_2 laser beam to seal blood vessels and lymphatics reduces the surgical complications

Figure 33a and b

(a) *Malignant melanoma of the flank*
(b) *The condition immediately after wide excision and split-thickness skin grafting. The sutures in this case were used to approximate the skin to the underlying deep fascia in order to obtain a more satisfactory cosmetic result. The graft, however, was not sutured in place but merely placed on the surgical defect*

Figure 34a and b

(a) *Malignant melanoma of the thigh*
(b) *The condition 4 months after wide excision and skin grafting. Note the perigraft injections of bacille Calmette – Guerin (BCG) vaccine as part of the immunologic treatment*

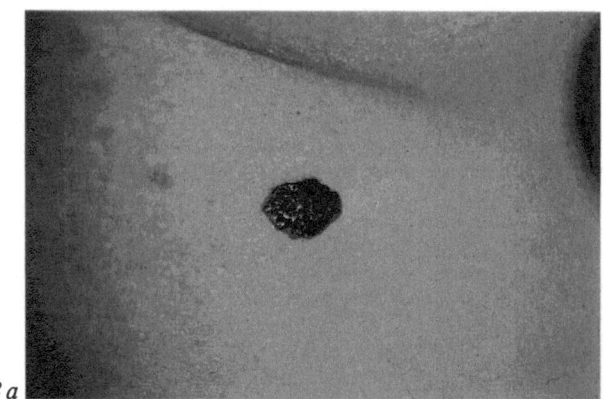

33 a

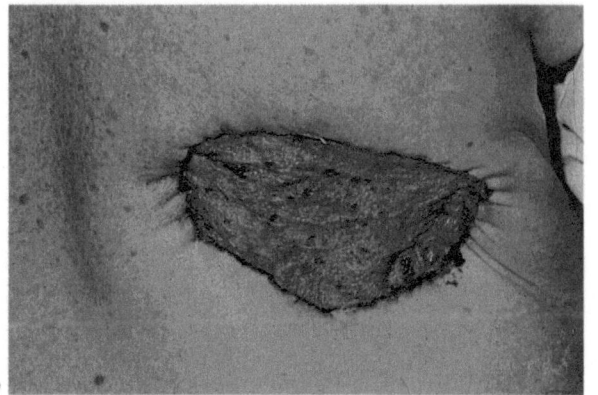

b

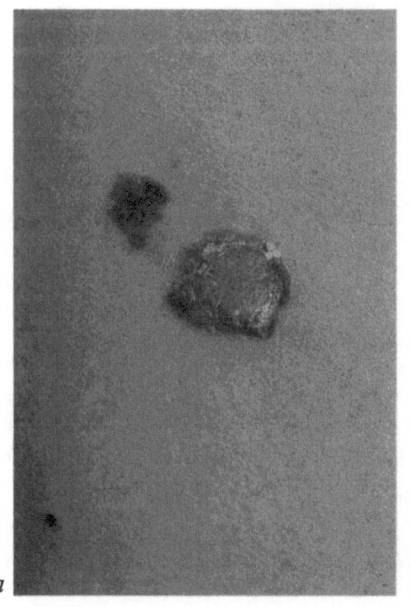

34 a

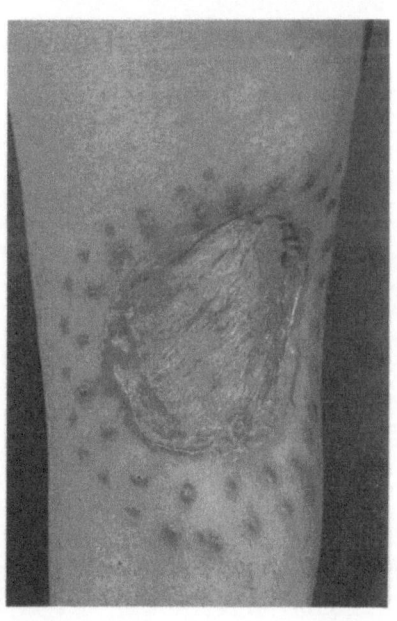

b

Figure 35 a–c

(a) Malignant melanoma of the inframammary region. In this case, wide excision and axillary lymph gland dissection were performed in conjunction. The surgical defect was partially sutured and partially covered with a split-thickness skin graft
(b) One week after the operation
(c) The condition 6 months after the operation. There is a contraction of the skin graft

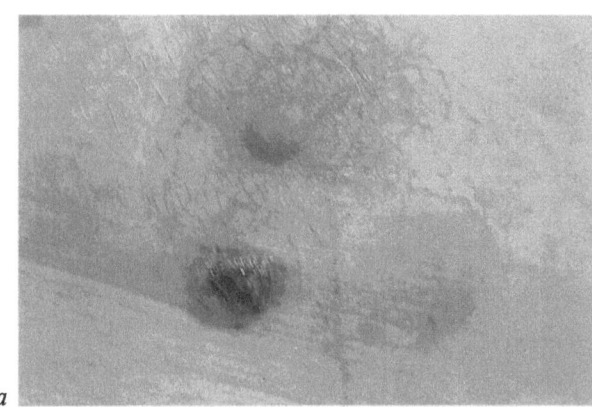

a

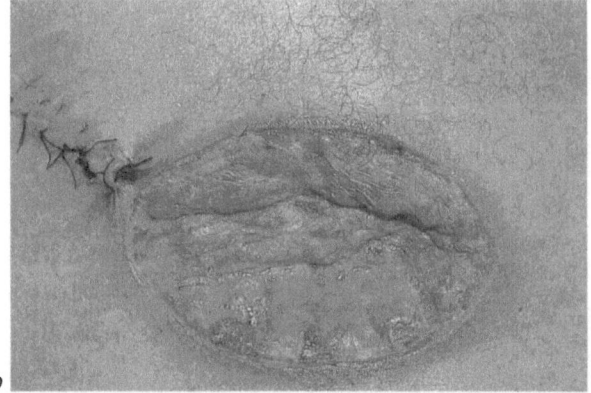

b

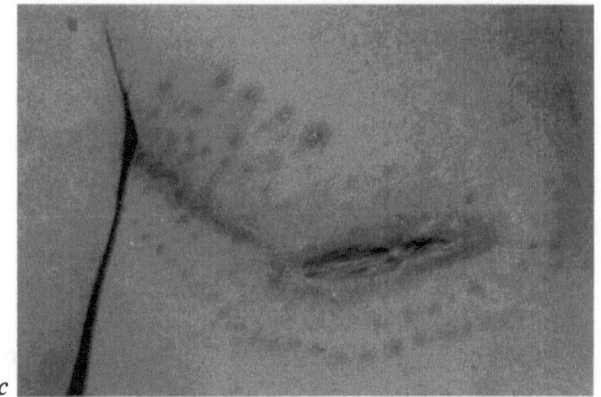

c

Figure 36a and b

(a) *Invasive malignant melanoma of the great toe. Using the CO_2 laser beam, amputation was performed including the head of the first metatarsal bone*
(b) *The condition 4 months after operation*

Figure 37a and b

(a) *Superficial spreading malignant melanoma of the thumb*
(b) *Two months after excision and split-thickness skin grafting. In this case, the depth of excision must be accurately controlled in order to assure the acceptance of the graft*

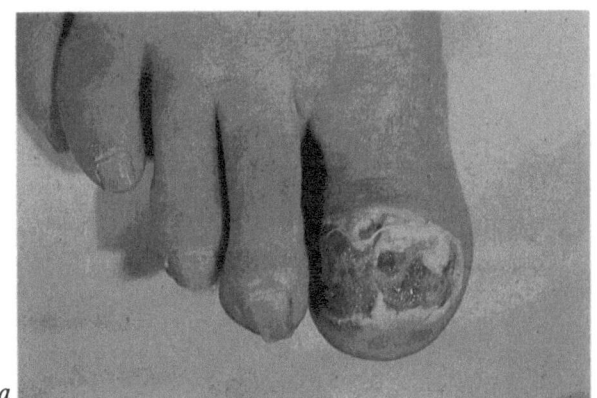

36 a

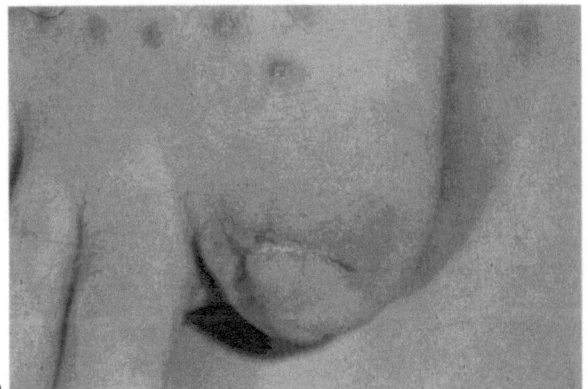

b

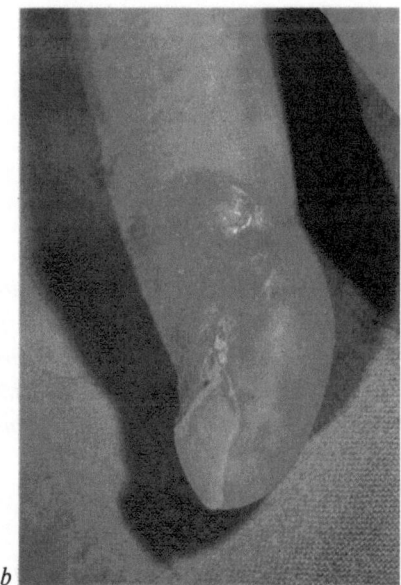

37 a

b

Figure 38a and b

(a) Inguinal lymph gland dissection for metastatic malignant melanoma. The incision is through the skin and fat tissue of the inguinal region
(b) The metastatic, black lymph glands are clearly demonstrated. In lymph gland dissection, the entire operation is performed with the CO_2 laser beam including the dissection around major blood vessels, thus, small blood vessels and lymphatics are also sealed

Figure 39

Another case of inguinal lymph gland dissection. The femoral artery is demonstrated

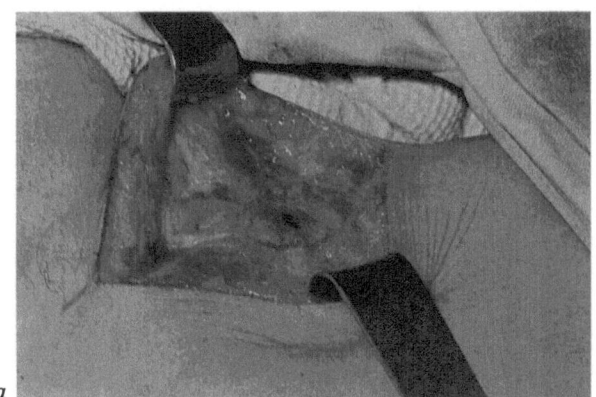

38 a

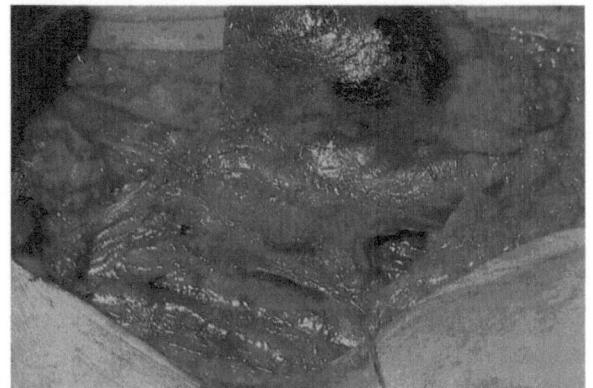

b

39

3.13.2 Other Malignant Tumors

Figure 40a and b

(a) Squamous cell carcinoma of the neck
(b) Two weeks after excision and primary suture

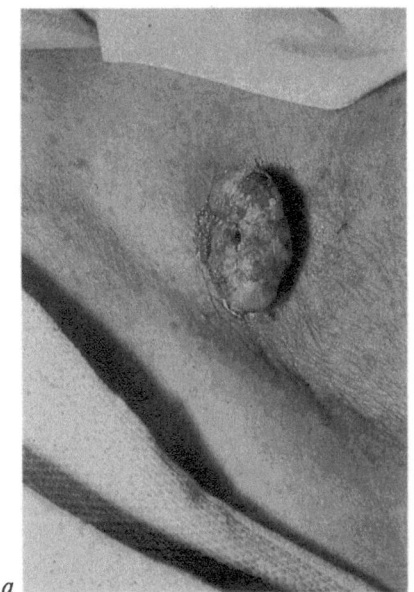

a

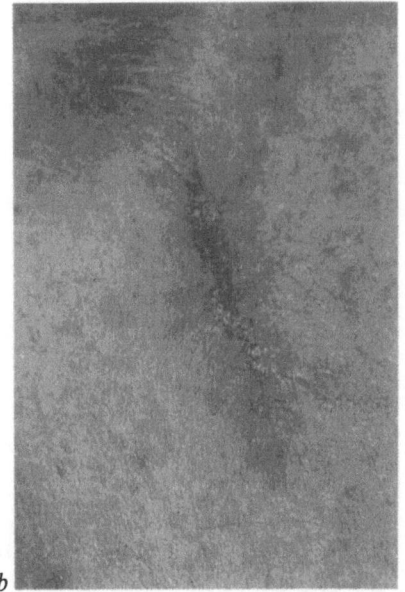

b

Figure 41a–d

(a) Recurrent fibrosarcoma of the back
(b) Twenty four hours after the operation, the split-thickness skin graft is placed on the surgical defect
(c) The condition 1 month after the operation
(d) The condition 6 months after the operation

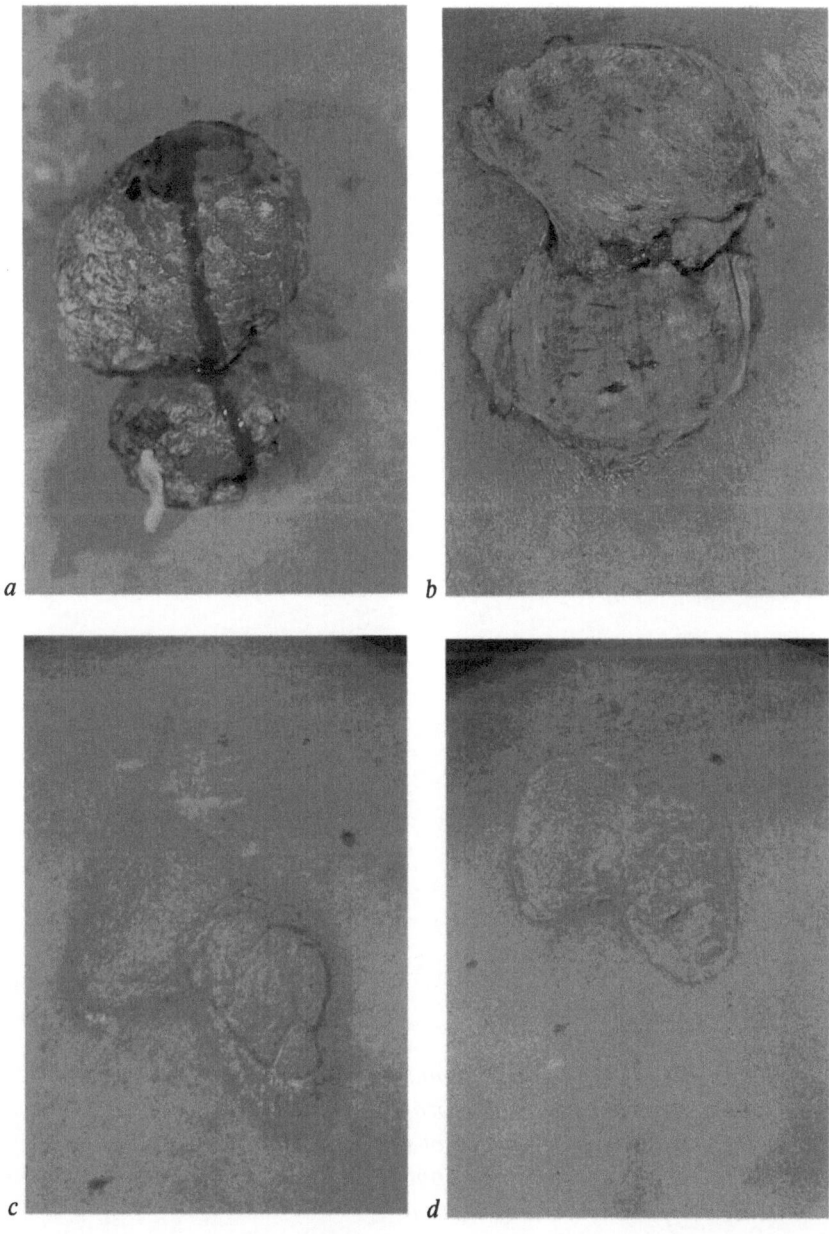

Figure 42a and b

(a) Pigmented basal cell carcinomas of the cheek and infraorbital regions and a small hemangioma above the eyebrow. Excision and primary suture of the basal cell carcinoma and vaporization of the hemangioma were performed

(b) Two months after the treatment there is only minimal scarring. Note in this case the patient was not shaven

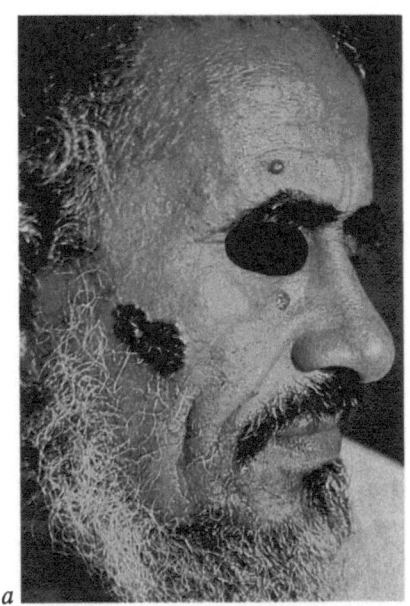

a

b

Figure 43 a–d

(a) Sweat gland carcinoma of the buttock. Wide excision and rotation flap were performed
(b) The condition 1 week after the operation
(c) The condition 1 month after the operation
(d) The condition 2 months after the operation

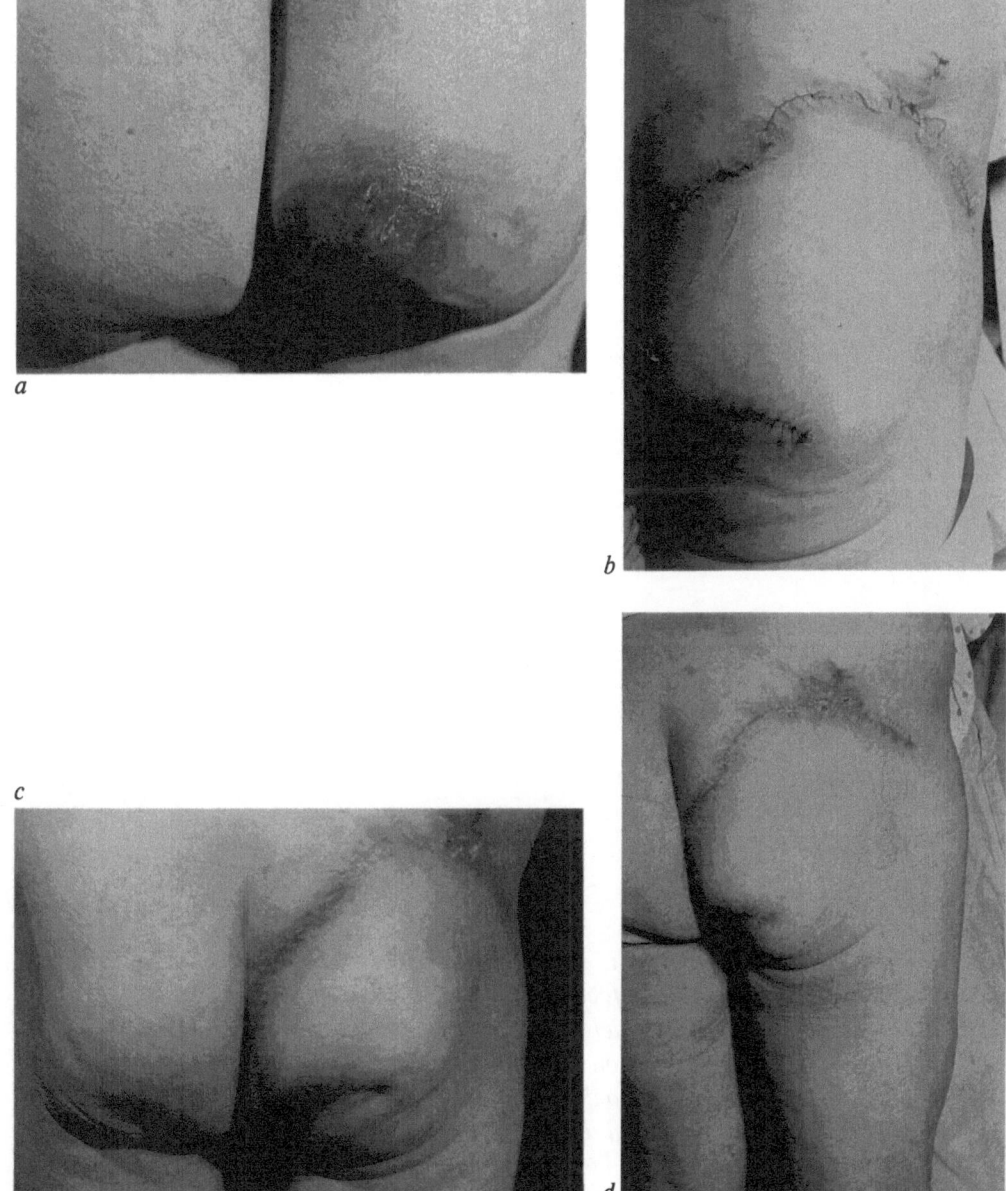

3.14 Operations Performed Through Highly Infected Tissues

Since the CO_2 laser seals blood vessels and lymphatics and at the same time sterilizes the tissue, it follows that operations can be performed through infected tissues without the fear of spreading the infection and thus precipitating septicemia.

Figure 44 a–d

(a) *A female child with synergistic gangrene of the vulva and perineum. The condition was severely toxic and the infection was spreading almost visibly*

(b) *Intraoperative view of the débridement performed with the CO_2 laser. In this case, an accurate visualization was possible due to the hemostatic effect of the laser beam, thus enabling us to distinguish between infected and noninfected tissues. The débridement was minimal and the operative bed dry and healthy*

(c) *The immediate postoperative view of the primary repair of the defect*

(d) *The condition 1 month after the operation*

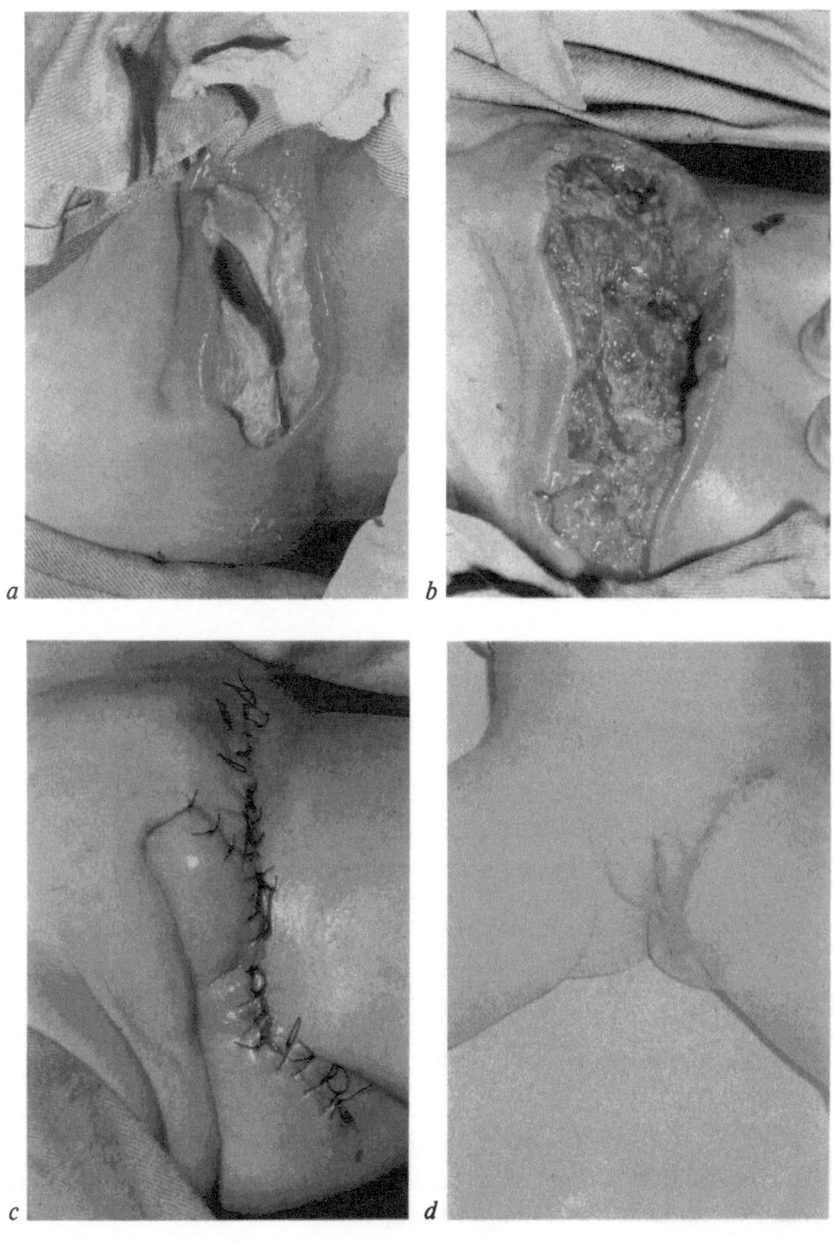

Figure 45 a and b

(a) *Synergistic gangrene of the leg in an elderly male. Débridement and split-thickness skin grafting were performed*
(b) *The condition 2 weeks after the operation*

Figure 46 a and b

(a) *An infected decubitus ulcer. Débridement and primary closure were performed*
(b) *The condition 3 weeks after the operation*

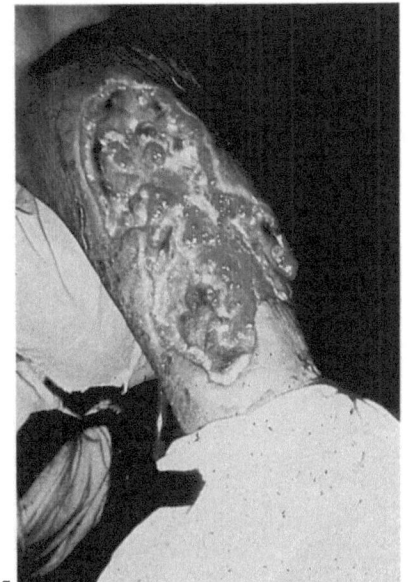
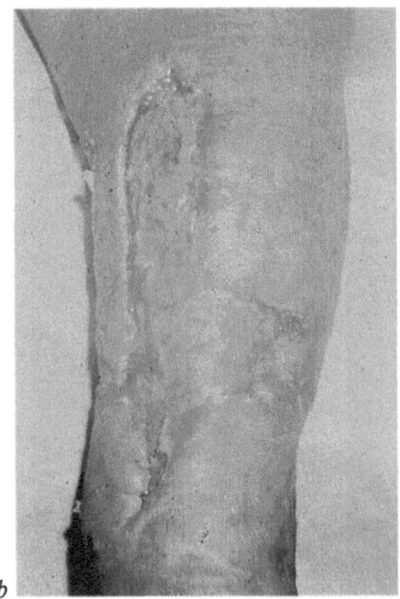

45 a　　　　　　*b*

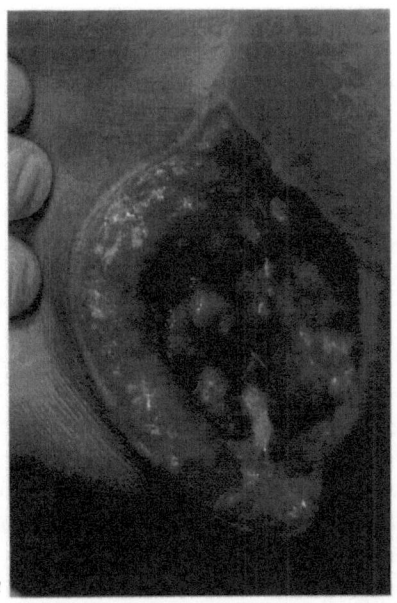
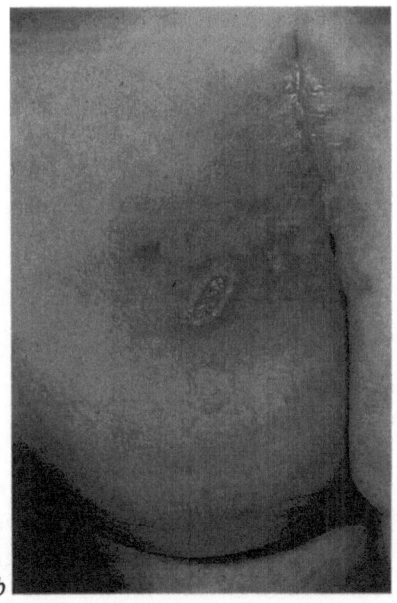

46 a　　　　　　*b*

Figure 47a–c

(a) *A typical infected recurrent pilonidal sinus. The pilonidal sinus was excised en bloc with margins of normal tissue down to the sacrococcygeal fascia*

(b) *The intraoperative view showing the defect; the black spots are sealed blood vessels. Primary closure in layers was performed*

(c) *The condition 1 month after the operation*

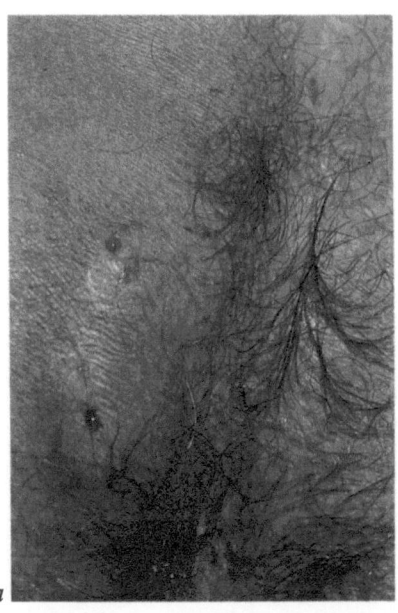

a

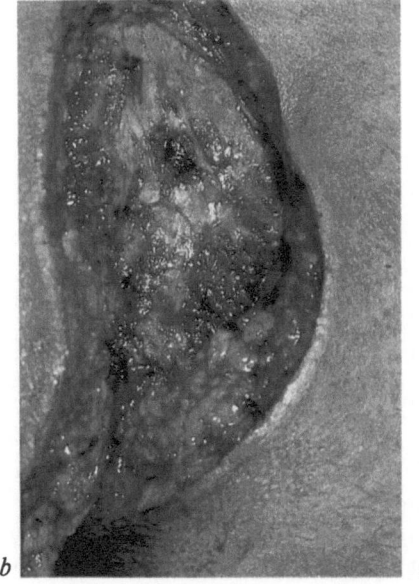

b

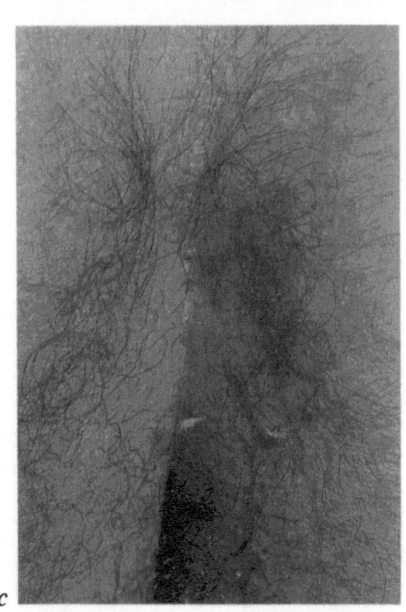

c

Figure 48 a–c

(a) Infected recurrent pilonidal sinus. In this case, en bloc excision was performed with split-thickness skin grafting
(b) The view a few days after the operation. The skin graft is held with sutures
(c) The condition 6 months after the operation

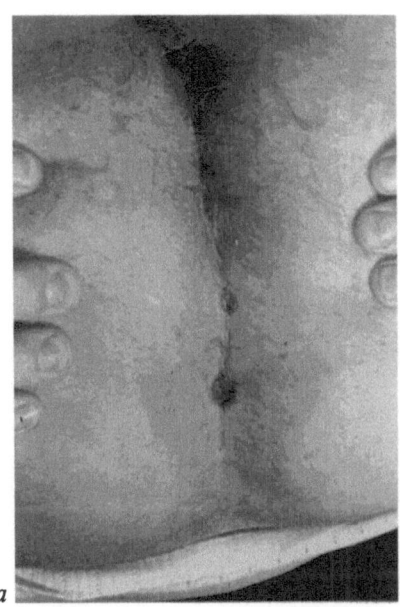

a

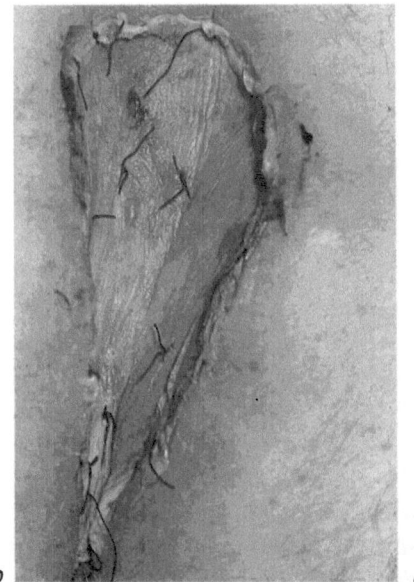

b

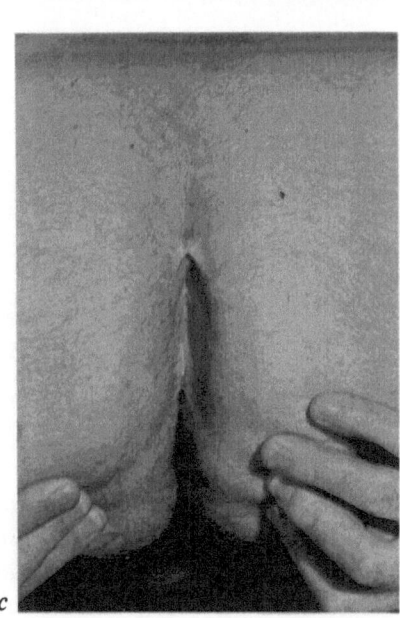

c

Figure 49a and b

(a) Multiple fistulae following an unsuccessful attempt at excising a pilonidal sinus. This patient also had a fibrosarcoma of the back

(b) The same patient Fig. 47a–e after excision and primary suture of the multiple fistulae and excision and grafting of the fibrosarcoma

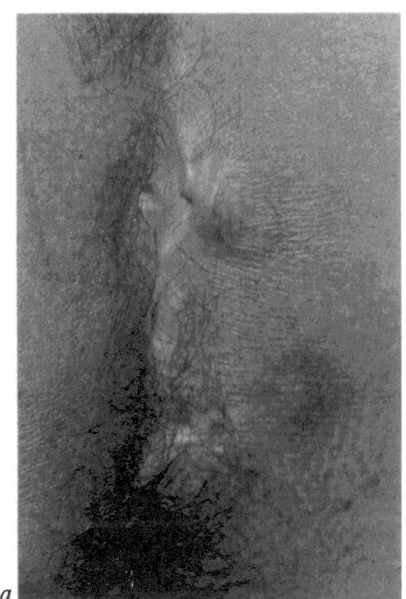

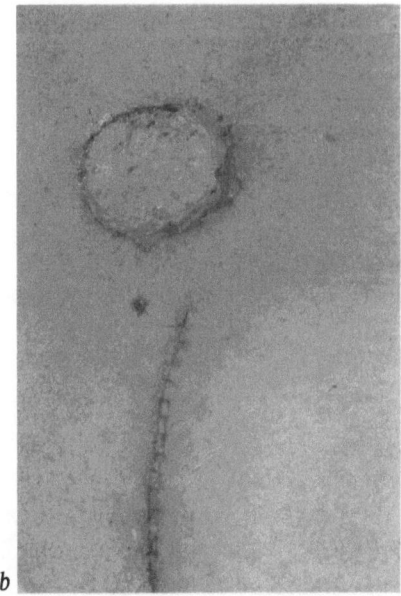

3.15 Other Conditions Treated with the CO_2 Laser

3.15.1 Neurofibromatosis

Figure 50a and b

(a) Large neurofibroma of the buttock. Excision and primary closure were performed
(b) The condition 3 months after the operation

3.15.2 Lipomata

Figure 51a–c

(a) Lipoma of the vulva. Excision and primary suture were performed
(b) The condition after 1 month. The scar is still erythematous
(c) The condition after 4 months

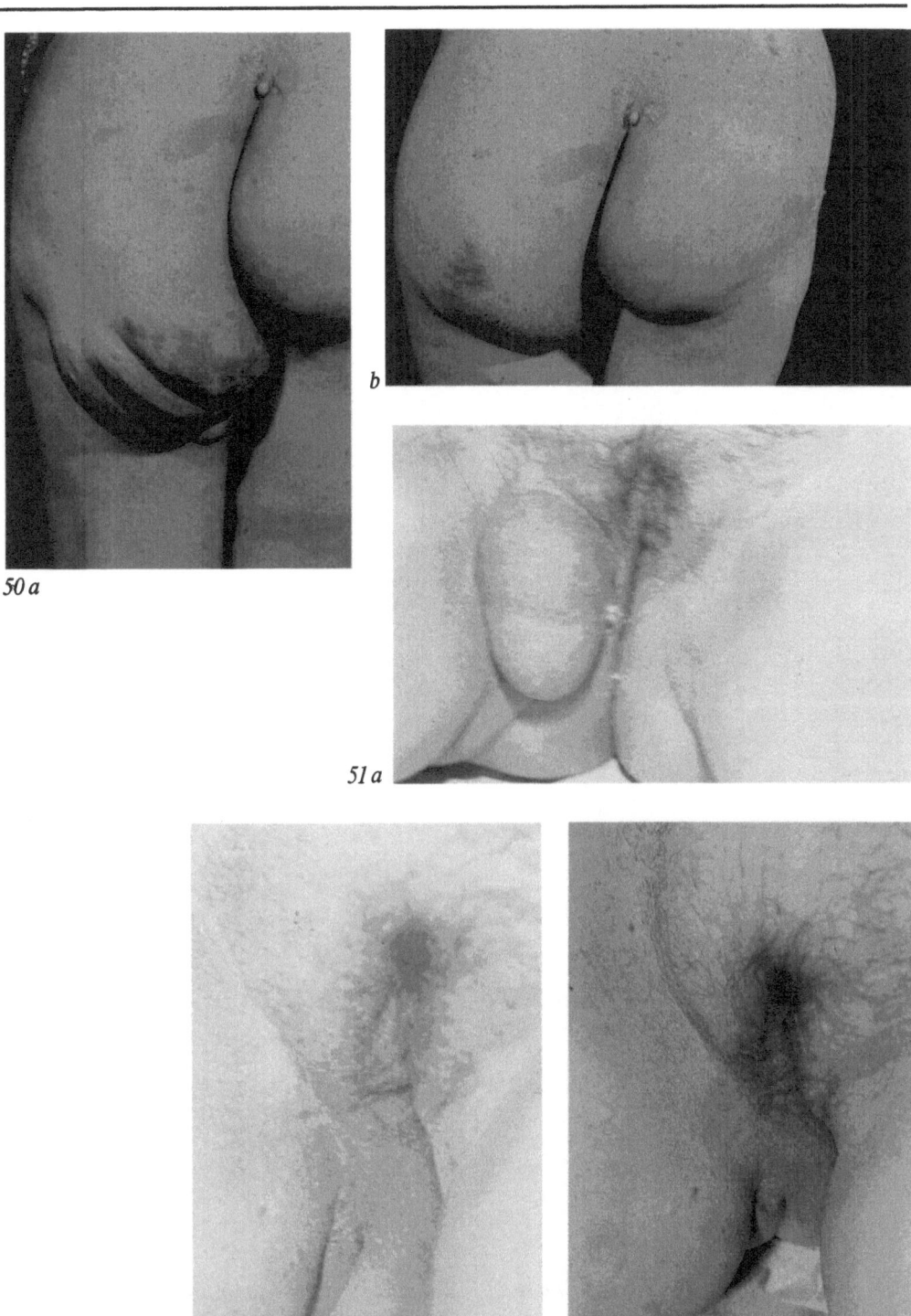

50 a

b

51 a

b

c

3.15.3 Leukoplakia

Figure 52a and b

(a) Leukoplakia of the vulva in an elderly woman. Wide excision and primary suture were performed

(b) The condition after 1 month

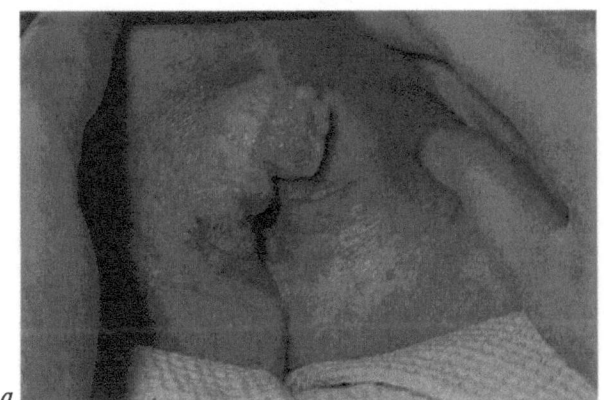

a

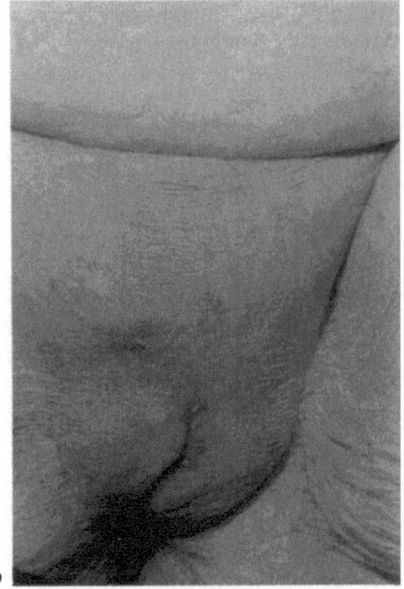

b

3.15.4 Condylomata Acuminata

Condylomata acuminata or venereal warts are common lesions of the genital, perineal, and anal regions. The infection is caused by a sexually transmitted strain of DNA-containing human papillary virus. The CO_2 laser beam is now a recommended modality for the treatment of these warts. The lesions can be excised using a focused beam or vaporized by a defocused beam. There is no intraoperative bleeding, less postoperative pain, prompt healing with minimal scarring and fibrosis and a lower recurrence rate than in other methods of therapy.

Figure 53a and b

(a) Condylomata acuminata of the scrotum. In this case, vaporization of the lesion was performed with secondary healing of the defect
(b) The condition after 2 weeks demonstrating the minimal amount of residual tissue damage

Figure 54a and b

(a) Circular condylomata of the anus. In this case, vaporization as well as excision were performed
(b) The condition 1 month after the operation

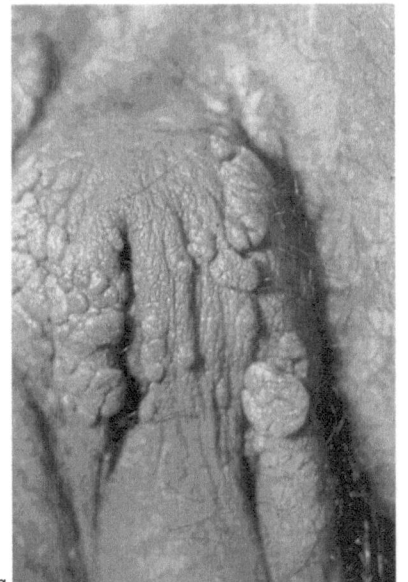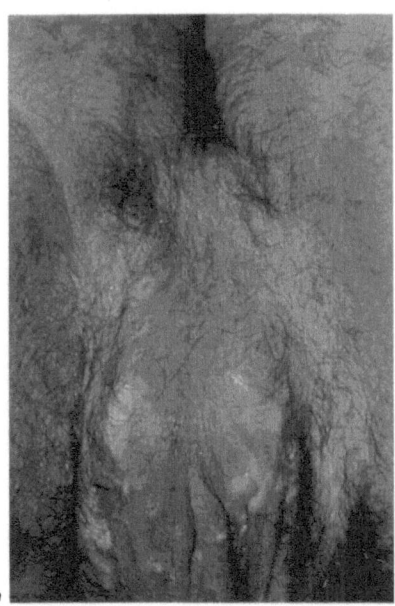

53 a   b

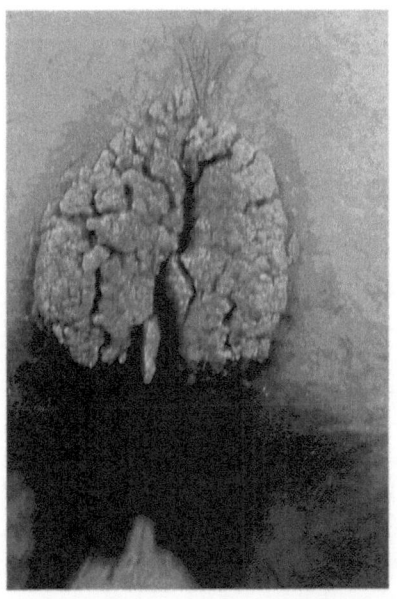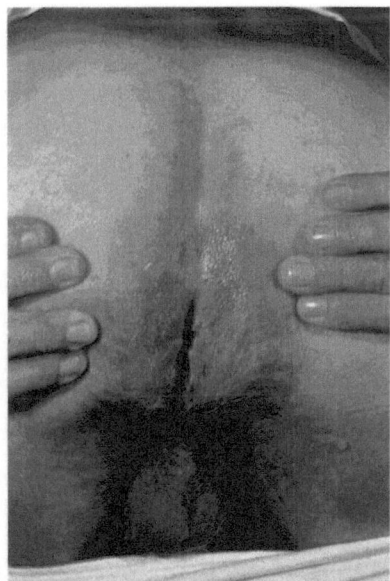

54 a   b

Figure 55 a–c

(a) *Recurrent condylomata of the anal region. The scars in the area indicate previous attempts at treatment*
(b) *The intraoperative condition*
(c) *One month postoperation*

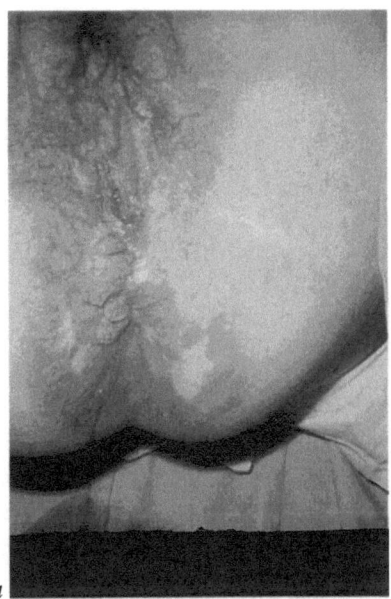

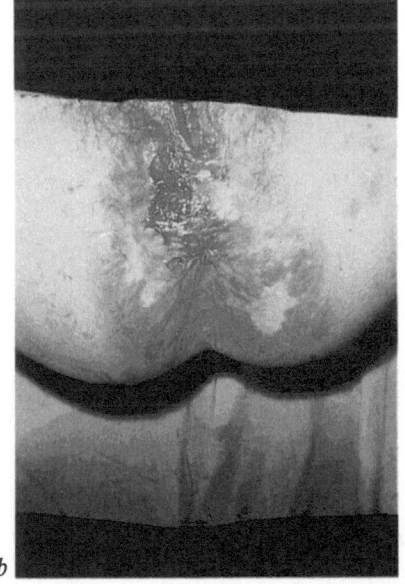

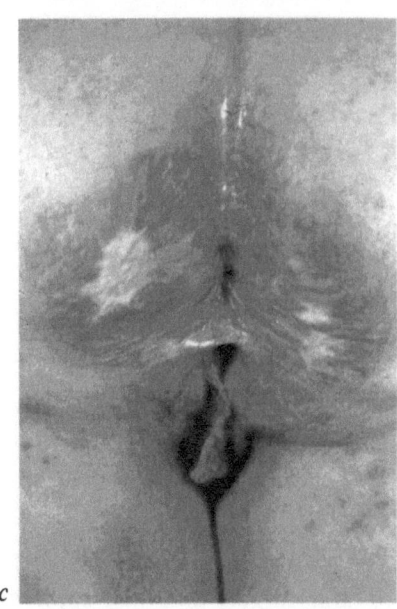

3.15.5 Warts

The use of the CO_2 laser is recommended in cases of warts since several reports indicate that the laser beam destroys the intracellular virus.

Figure 56a–c

(a) *Massive warts on the sole of the foot of a young girl. Various therapeutic modalities were used for the treatment of this condition without any success. The CO_2 laser was used to excise the massive growth*
(b) *The intraoperative view. Split-thickness skin grafting was used to cover the defect*
(c) *The condition 4 months after the operation*

3.15.6 Cavitational Surgery

Figure 57

The double-puncture laparoscope in use with the Sharplan 733 laser for the purpose of tubal surgery and the treatment of such intra-abdominal conditions as endometriosis

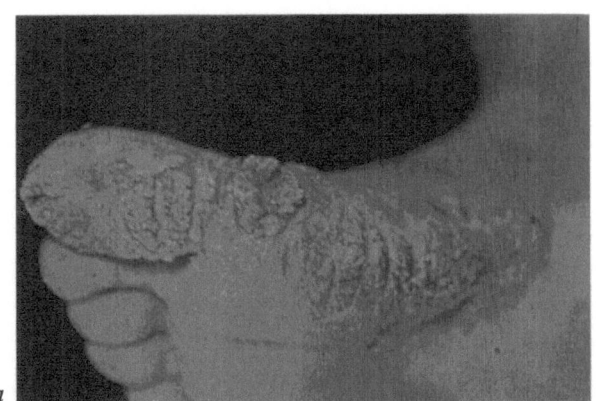

56 a

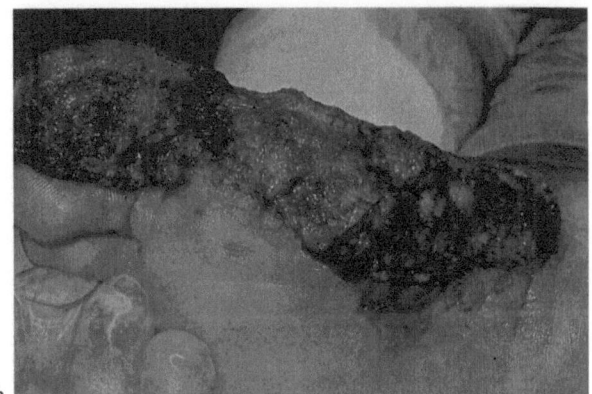

b

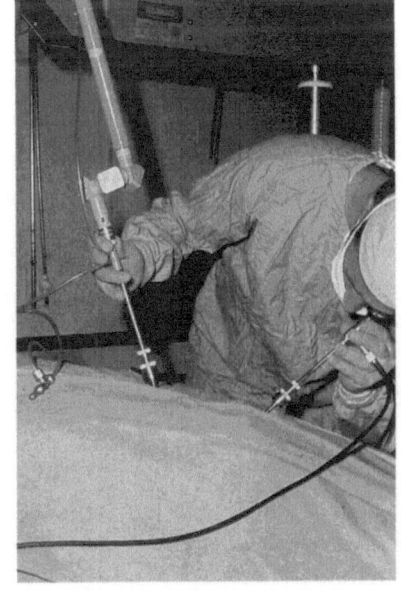

57

c

Figure 58a and b

(a) Adenocarcinoma of the rectum where radical surgery was contraindicted because of the general condition of the patient
(b) The view after laser excision of the tumor and vaporization of the base

Figure 59a and b

(a) A polyp of the vocal cord
(b) The condition immediately after laser vaporization with the aid of a micromanipulator

58 a

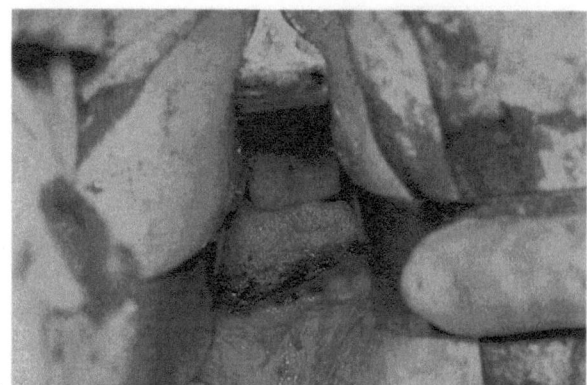

b

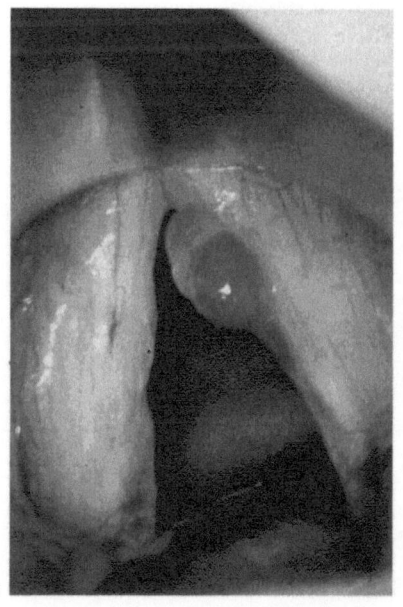

59 a

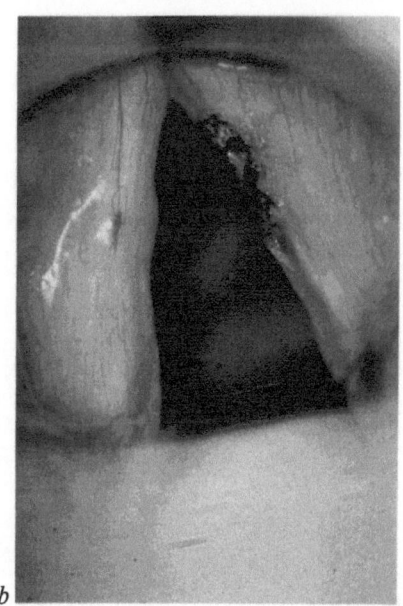

b

Part B

The CO_2 Laser in Dermatologic Surgery

In recent years, we have used the CO_2 laser for the ambulatory treatment of various cutaneous lesions. The following is a classified list of lesions diagnosed clinically or histologically and treated accordingly.

1. Nevi	Epithelial
	Papillomatous
	Verrucous

2. Vascular lesions	Hemangiomata
	Nevus araneus
	Senile angiomata
	Telangiectasia
	Angiokeratoma
	Pyogenic granuloma
	Glomus tumors

3. Other lesions	Papillomatosis	Atheromata (sebaceous cysts)
	Skin tags	Decorative tattoos
	Senile keratosis	Accessory auricles
	Seborrheic keratosis	Accessary fingers
	Cornu cutaneum	Superficial basal cell carcinomata
	Verrucae	Kaposi's sarcomata
	Condylomata acuminata	Bowen's disease
	Keratoacanthoma	Juvenile melanomata
	Leukoplakia	Xeroderma pigmentosa
	Xanthelasma	Xanthomata
	Fibromata	
	Neurofibromatosis	

In dermatologic surgery, the laser beam is used for the excision or vaporization of lesions. The procedure, which is generally conducted in an outpatient clinic is rapid and uncomplicated, bloodless, sterile, almost painless, and well tolerated by children and the elderly. Postoperatively, there is no discomfort for the patient and no dressings are required; postoperative pain and infective complications do not occur. The treated areas heal rapidly because the skin appendages escape permanent damage. The cosmetic results of the treated areas are superior to those obtained by other methods and there is only minimal scarring.

1 Method of Treatment

All procedures are performed ambulatorily in our laser outpatient clinic using the Sharplan CO_2 surgical laser. There is no need to prepare the skin because of the sterilizing properties of the laser beam. The surgical procedures are performed with single pulses of the laser beam with a time exposure of $1/20$ s and $10-20$-W output, so that in almost half of the cases no local anesthesia is required since the short duration of the pulses makes the procedure almost painless. In cases where local anesthesia is required (1% lidocaine without epinephrine), the exposure time is $^1/_2$ s, or continuous exposure is performed. The treatment involves, excision (Figs. 60a–c) or vaporization of the lesions. When excision has been performed, a histologic examination can be carried out.

After the treatment, the resulting charred tissue is removed with a swab moistened with saline. The wounds are not dressed and the patients are advised to wash the treated areas with soap and water twice a day.

The excision or vaporization of a lesion (Fig. 61a) causes a shallow crater surrounded by slight erythema (Fig. 61b). In cases where excision has been performed, the lesion is sent for histologic examination (Fig. 61c). Following treatment, there is exudation and crust formation (Fig. 61d), its thickness depending on the depth of the necrotic zone. Subsequently, the crust separates and re-epithelialization and complete healing is achieved in 2–4 weeks, depending on the size and depth of the crater. A pink soft scar is left (Fig. 61e), which gradually fades over a period of $1-2$ months and becomes indistinguishable from the surrounding skin (Fig. 61f).

It is worthy of note that after treating thousands of lesions with the above technique, there were only a few cases of postoperative infective complications and hypertrophic scarring (less than 0.5%), even in those areas of the body where hypertrophic scarring usually occurs. In several cases, hypopigmentation, occurred in the treated areas, mainly on the upper back. Subsequently, the hypopigmentation disappeared after several months. Hyperpigmentation did not occur in any of the patients treated. Only a few scars were slightly depressed due to deeper penetration of the laser beam in larger lesions. However, after several months the depressed scars became flat.

Figure 60

(a) A neurofibroma about to be excised
(b) Beginning of the excision using a pulsed mode
(c) At completion of excision, note shallow crater

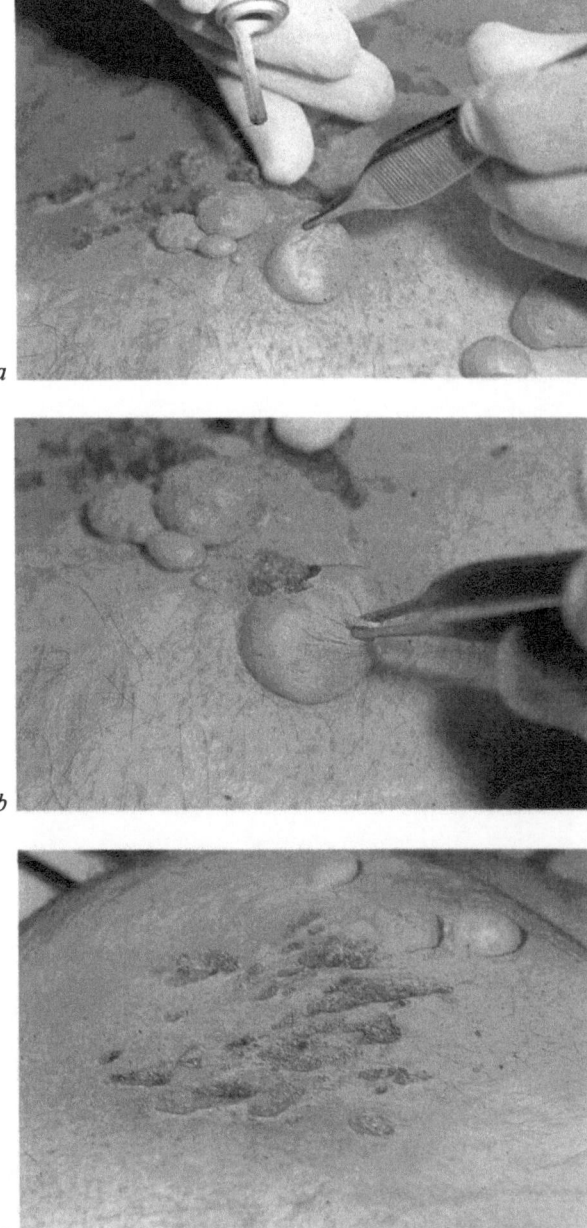

a

b

c

Figure 61 a–f

(a) Hemangioma of the upper lateral aspect of the neck in a 44-year-old female
(b) The immediate posttreatment results showing a shallow crater surrounded by slight erythema
(c) The histology of the excised lesion showing a cavernous hemangioma

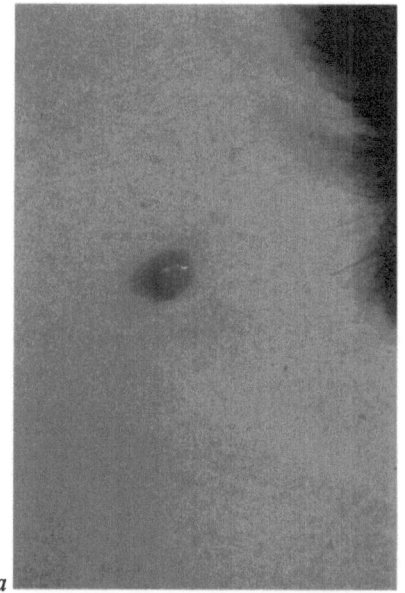

a

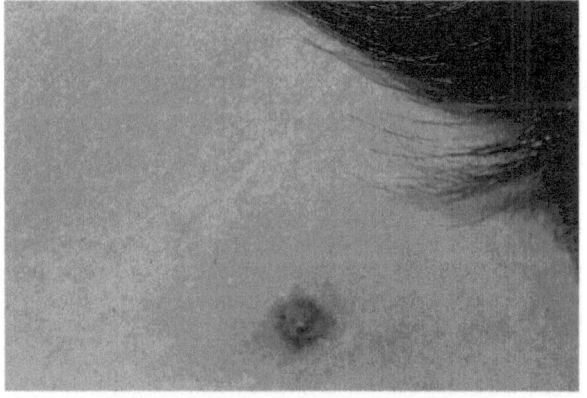

b

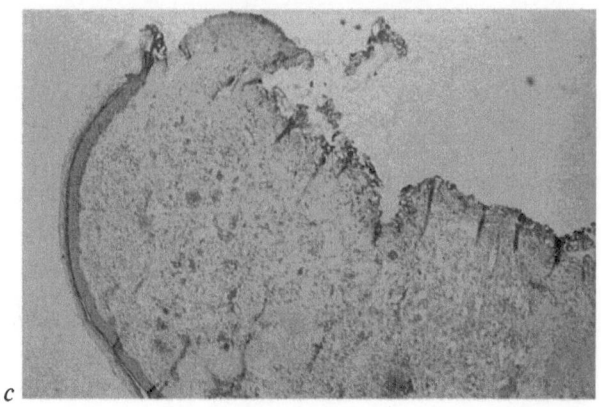

c

Figure 61

(d) Ten days after treatment there is a scab which partially covers the charred surface
(e) Two weeks after treatment – a soft pink scar
(f) One month after treatment, the scar is almost indistinguishable from normal skin

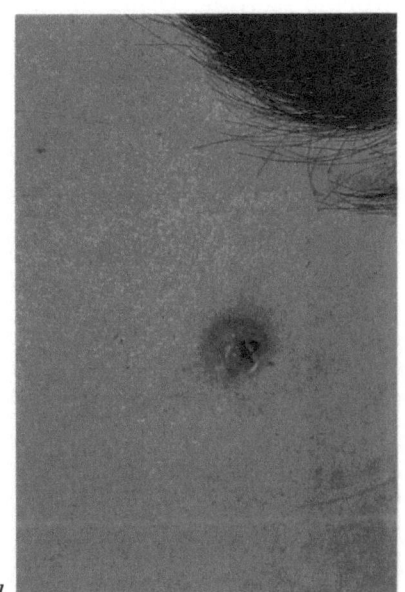

d

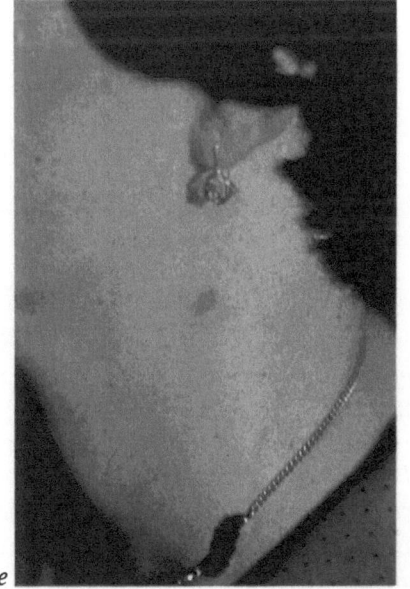

e

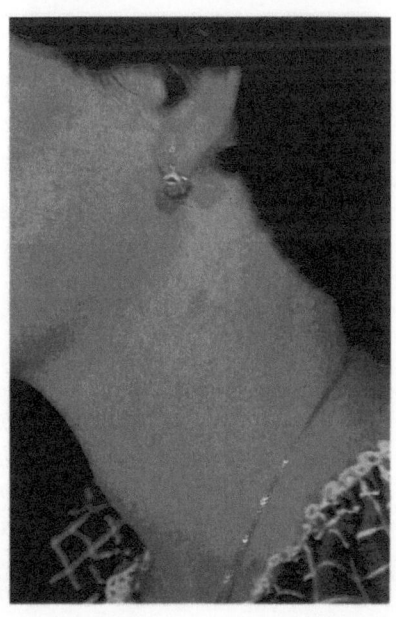

f

1.1 Removal of Decorative Tattoos

There are a number of methods for the treatment of decorative tattoos including excision with primary closure or skin grafting, dermabrasion, electrocautery, or cryosurgery. However, all these methods are impaired by several disadvantages, such as incomplete removal of the pigmented areas, postoperative pain, or hypertrophic scarring. Using the CO_2 laser for the removal of decorative tattoos, we can control the precise depth of destruction of the pigmented areas. There is less postoperative pain and less scarring.

Figure 62a–f

(a) *Pretreatment view of a tattoo of the arm*

(b) *Laser dermabrasion is performed at low power (5 W) in order to visualize the pigment more accurately. The tattoo is selectively vaporized down to the dermal level using local anesthesia and a continuous laser beam*

(c) *The laser-treated layer is removed by wiping with a moist pad so that the remaining pigment becomes more obvious. Following this procedure, the residual pigment is removed by vaporization in the pulsed mode*

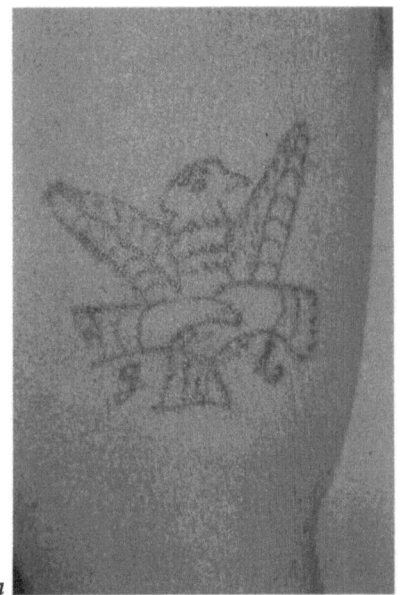

a

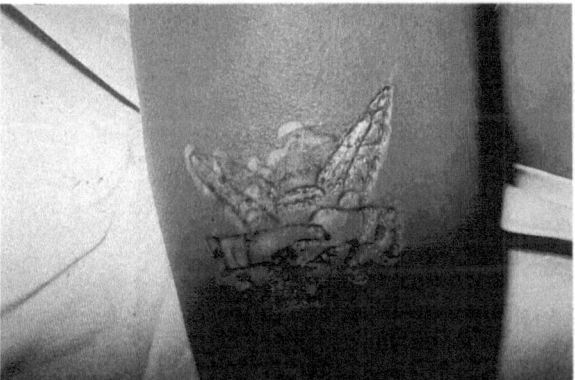

b

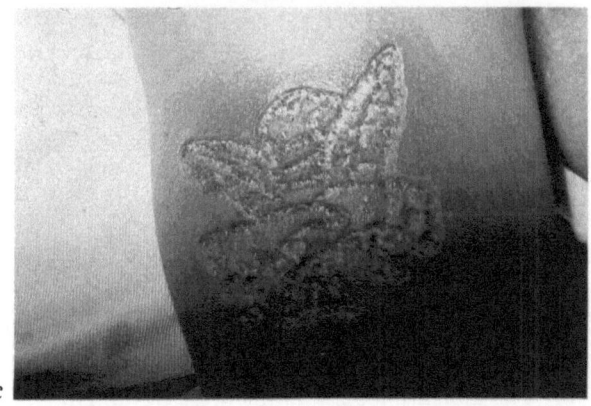

c

Figure 62

(d) *The immediate posttreatment appearance. A Bethamethasone cream is used for dressing in order to minimize the reaction of the surrounding tissues and to protect the wound from postoperative inflammation and thus decrease scar formation*

(e) *The appearance 2 weeks after treatment*

(f) *The appearance 2 months after treatment*

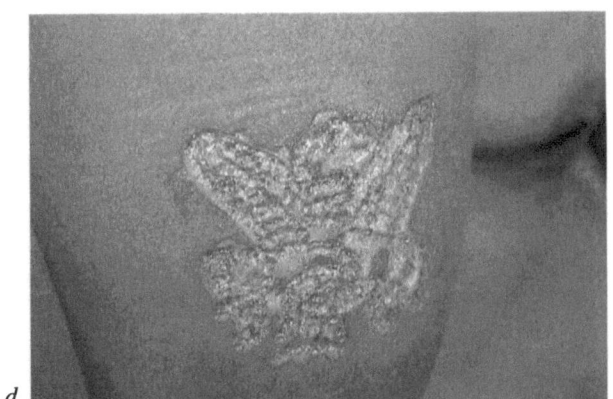

d

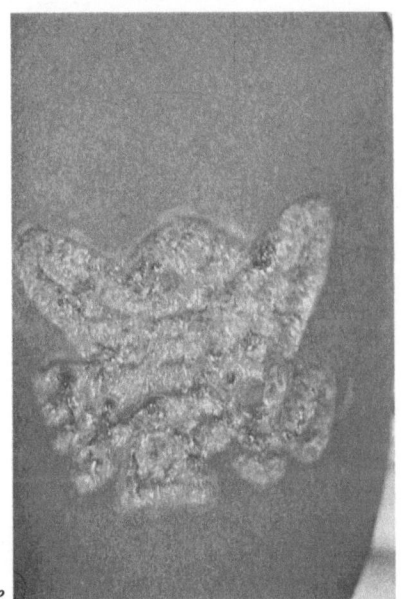

e

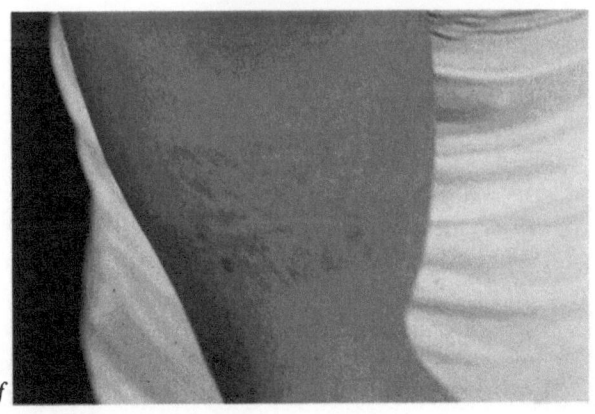

f

Figure 63 a–c

(a) *Pretreatment view of a different tattoo on the forearm*
(b) *The immediate postoperative appearance. The black spots are carbonized remnants that are wiped away leaving a dry wound*
(c) *The appearance 5 months later. The scars show mild hypopigmentation with preservation of skin texture, contour and hair follicles. In cases where residual pigment is still present after a single session, treatment is repeated 1 month later*

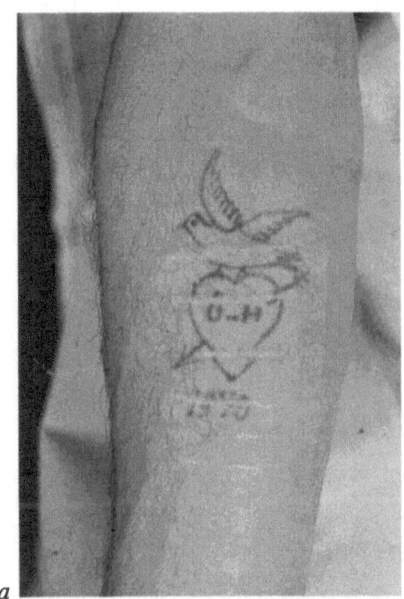

a

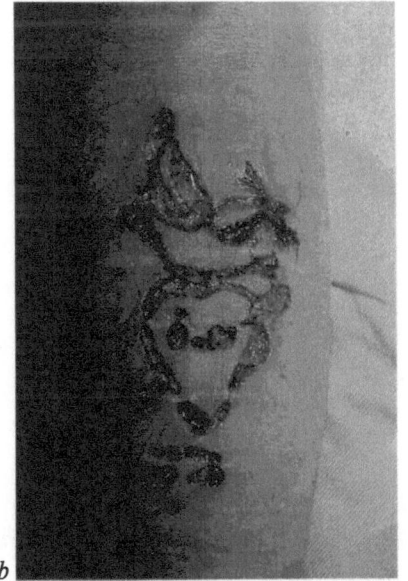

b

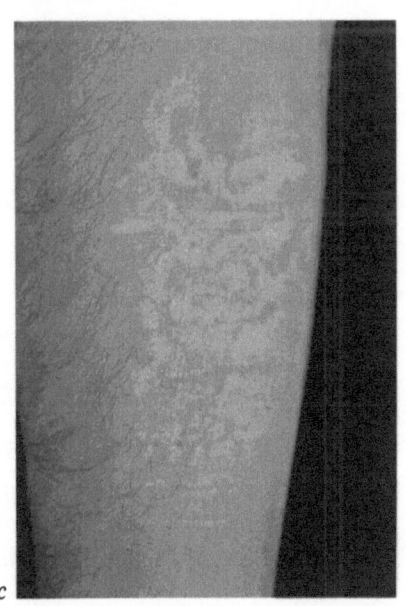

c

2 Superficial Vascular Lesions

2.1 Telangiectasia

One of the most important attributes of the CO_2 laser beam is its ability to seal off small dilated dermal blood vessels through intact skin, a condition which would otherwise be untreatable especially on the face. The blood vessels and their ramifications are sealed with a focused beam using single shots. The pulses are continued until the telangiectasis disappears. It is worthy of note that in the event of bleeding, pressure should be applied and no attempt made to coagulate the bleeding points.

This method was also used successfully in cases of familial teleangiectasia of the nasal mucosa (Rendu-Osler-Weber syndrome). This syndrome, which is inherited as an autosomal dominant, is characterized by the familial occurrence of numerous telangiectases on the skin and mucous membranes and repeated episodes of haemorrhage. The syndrome is very rare and is commonest among Jews. Although the lesions may involve any structure of the body, the commonest symptom is recurrent epistaxis since the mucous membranes of the nasal septum are most frequently involved.

Figure 64 a–c

(a) Telangiectasia on the tip of the nose

(b) The immediate posttreatment condition showing minute pits and complete obliteration of blood vessels

(c) One week after treatment, the telangiectasia has disappeared without visible scarring

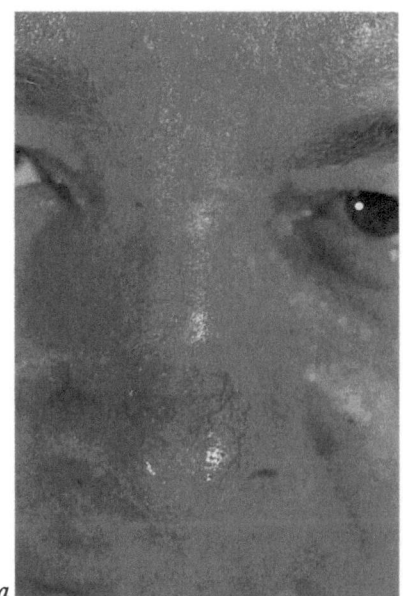

a

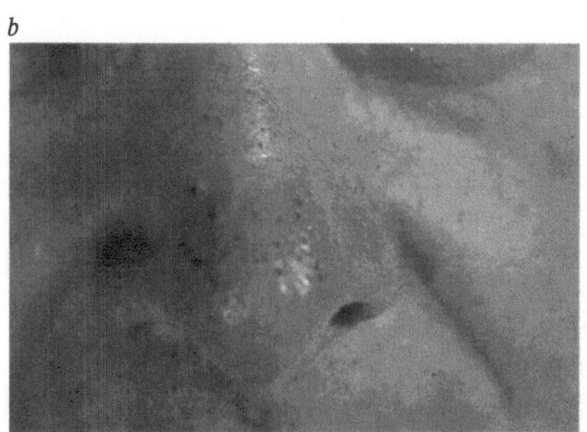

b

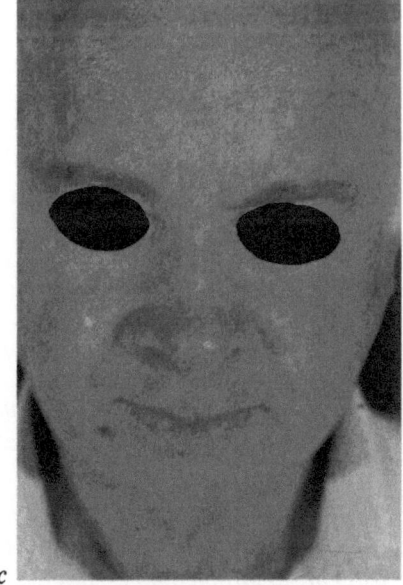

c

Figure 65 a–d

(a) Telangiectasis of nose and chin
(b) Telangiectasis of nose and chin. Effective destruction of the blood vessels was achieved using single pulses of the laser beam radiated to the central vessels and along the routes of the blood vessels
(c) The condition 1 week later
(d) The condition 1 week later

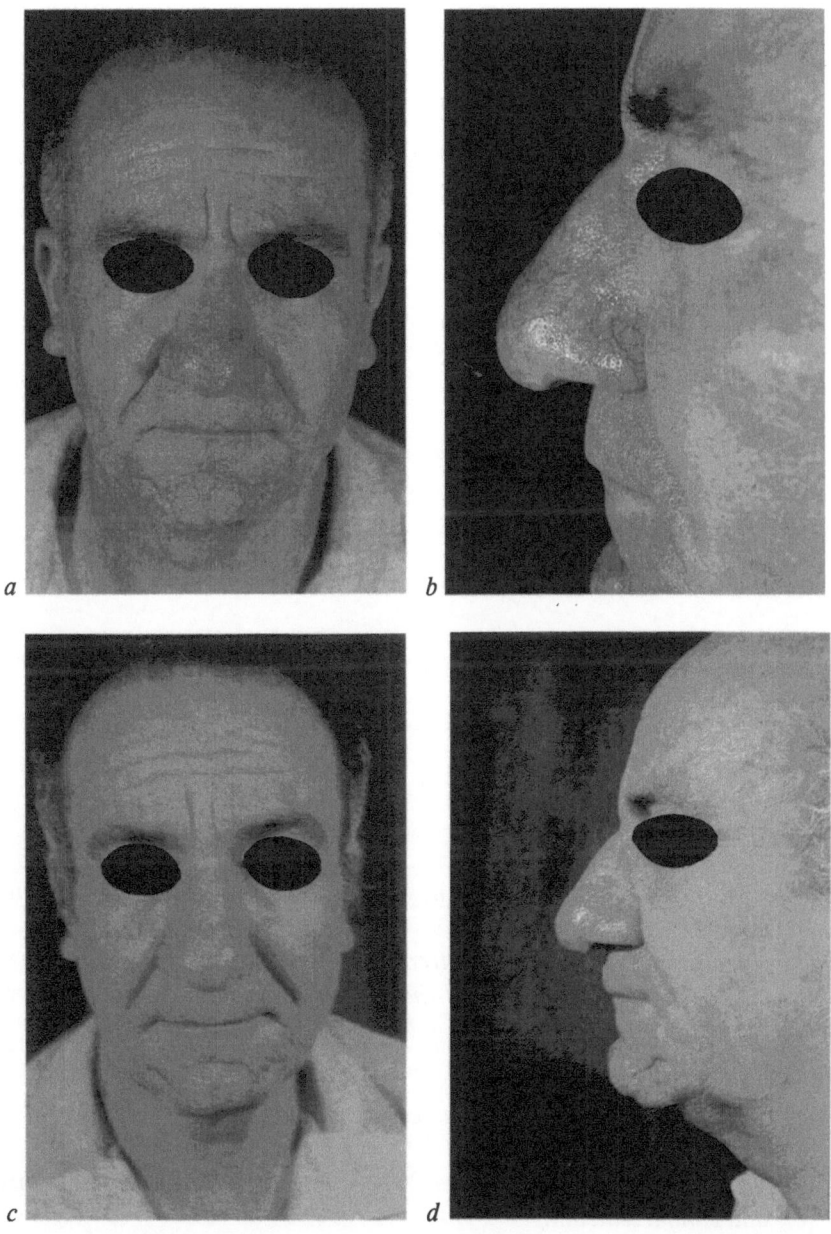

a

b

c

d

2.2 Cavernous Hemangioma

Figure 66a and b

(a) *Cavernous hemangioma of nose in a 7-year-old boy. The lesion was vaporized under local anesthesia*

(b) *Appearance 2 months after treatment. When using the pulsed laser beam in short bursts, there is only minimal thermal damage and thus only minimal scarring*

Figure 67a and b

(a) *A 32-year-old female with a venous lake on the lower lip. In this case, the lesion was vaporized*

(b) *The condition 1 month later. There is a complete disappearance of the lesion and complete healing*

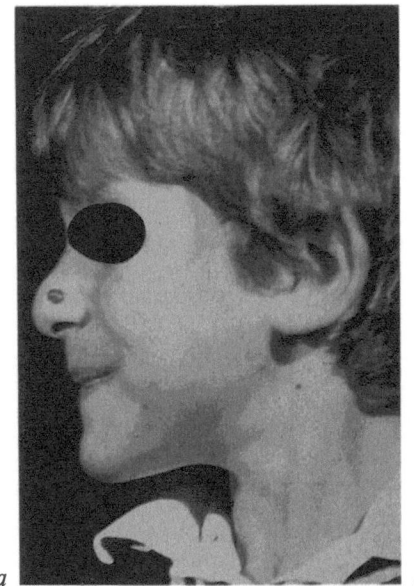 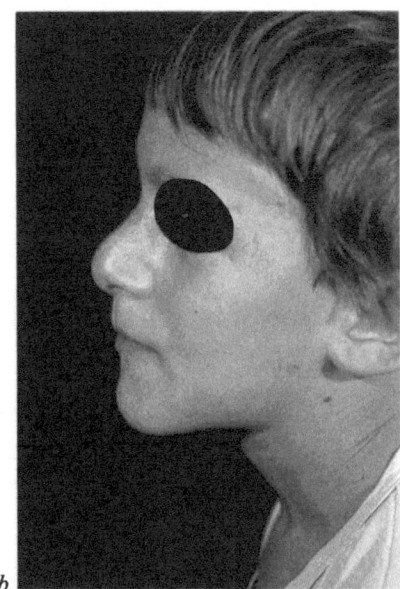

66 a   *b*

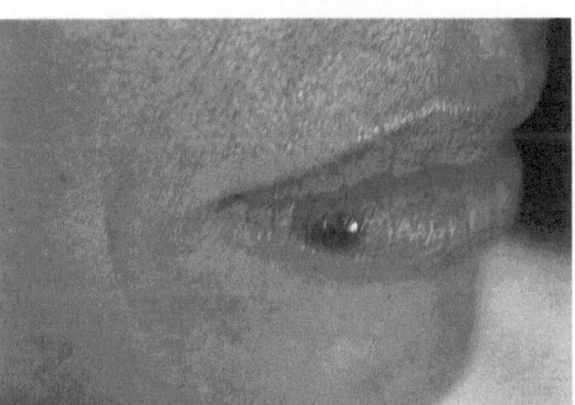

67 a

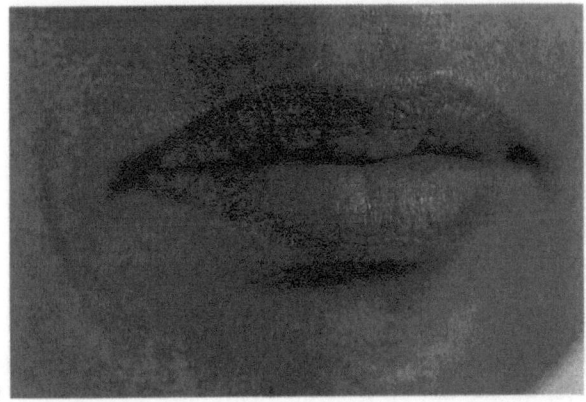

b

Figure 68a and b

(a) Pretreatment view of a hemangioma on the shoulder of an 8-year-old child
(b) One month after treatment, only a minimal pink scar exists

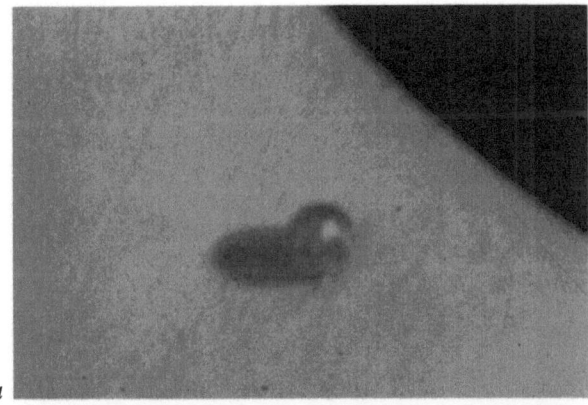

a

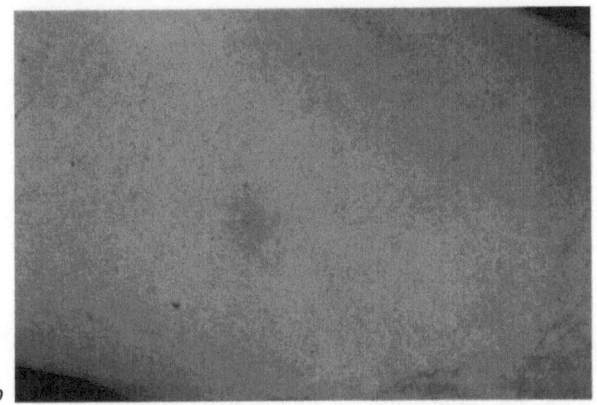

b

2.3 Senile Angiomas (De morgan's spots)

Figure 69a–c

(a) A 67-year-old male with multiple senile angiomas on the chest and abdomen
(b) Immediate postoperative appearance
(c) The appearance 1 month after treatment

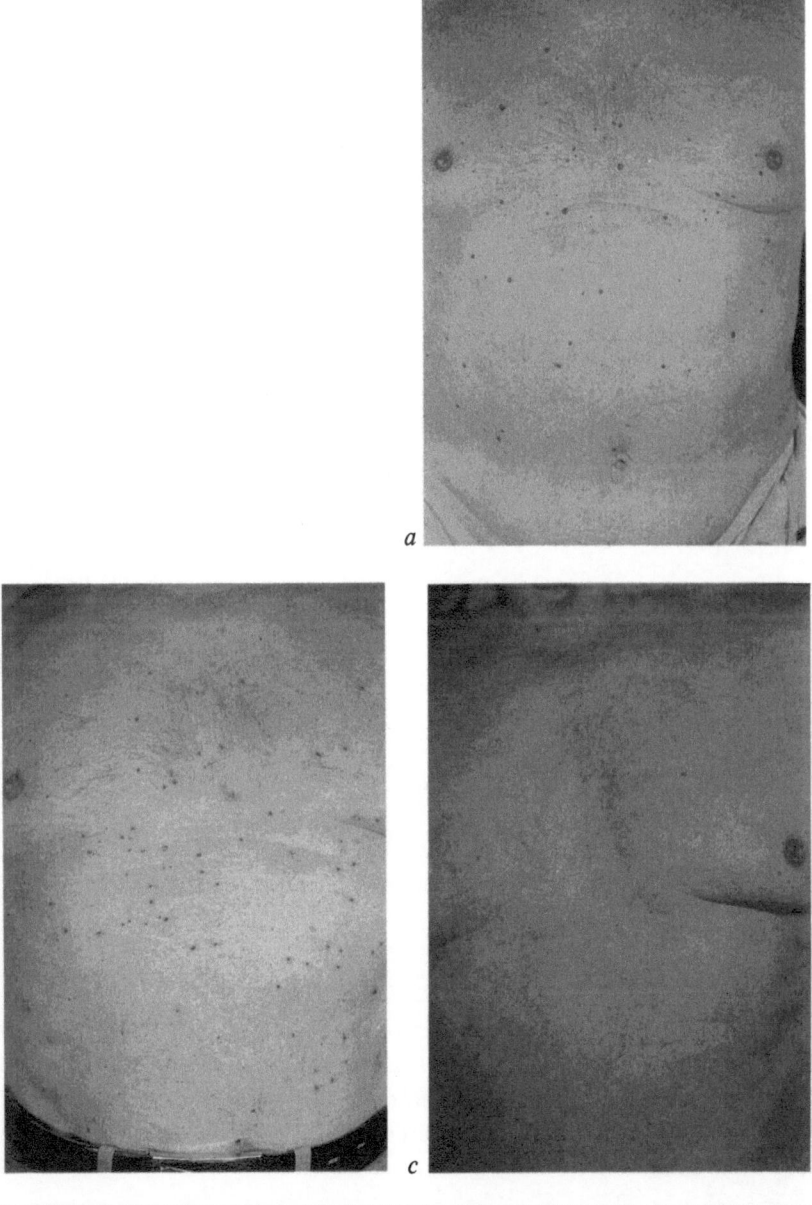

2.4 Strawberry Nevus

These are the commonest type of cutaneous hemangioma and may occur anywhere on the cutaneous surface. Ordinarily, complete spontaneous regression is expected, but in some cases massive bleeding may occur and thus treatment is required.

Figure 70 a and b

(a) A bleeding strawberry nevus of the bridge of the nose in a 4-month-old male child. To protect the eyes, the orbital areas were covered with wet gauze. The procedure was performed without local anesthesia, vaporization was performed, and only a few laser shots were necessary for the disappearance of the nevus

(b) The condition 1 month after treatment

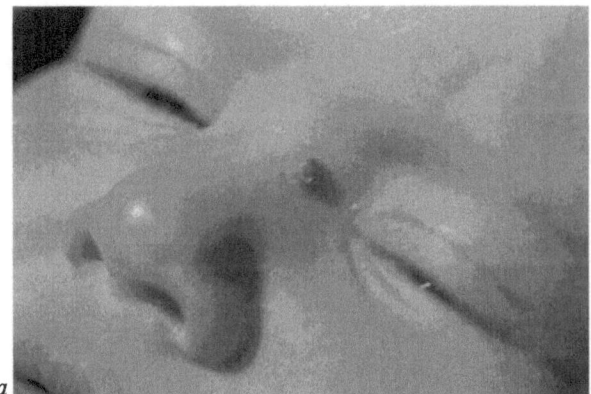

a

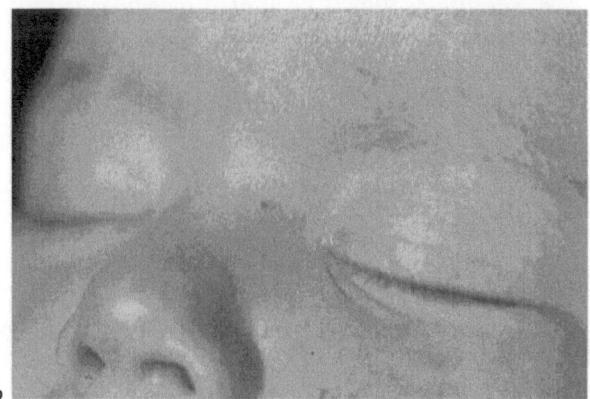

b

2.5 Angiokeratoma

Figure 71 a and b

(a) The pretreatment view of a bleeding angiokeratosis of the scrotum. The lesions were vaporized with a defocused beam

(b) The appearance 1 month after treatment. In some cases more than one treatment is necessary for the removal of the lesions

Figure 72 a–c

(a) The pretreatment view of a bleeding angiokeratosis of the thigh in a 32-year-old woman. A biopsy was taken with a focused laser beam and then the lesions were vaporized with a defocussed beam

(b) The appearance 2 weeks after treatment. Note that scars are still present in some areas

(c) The appearance 1 month after treatment. Some lesions still exist

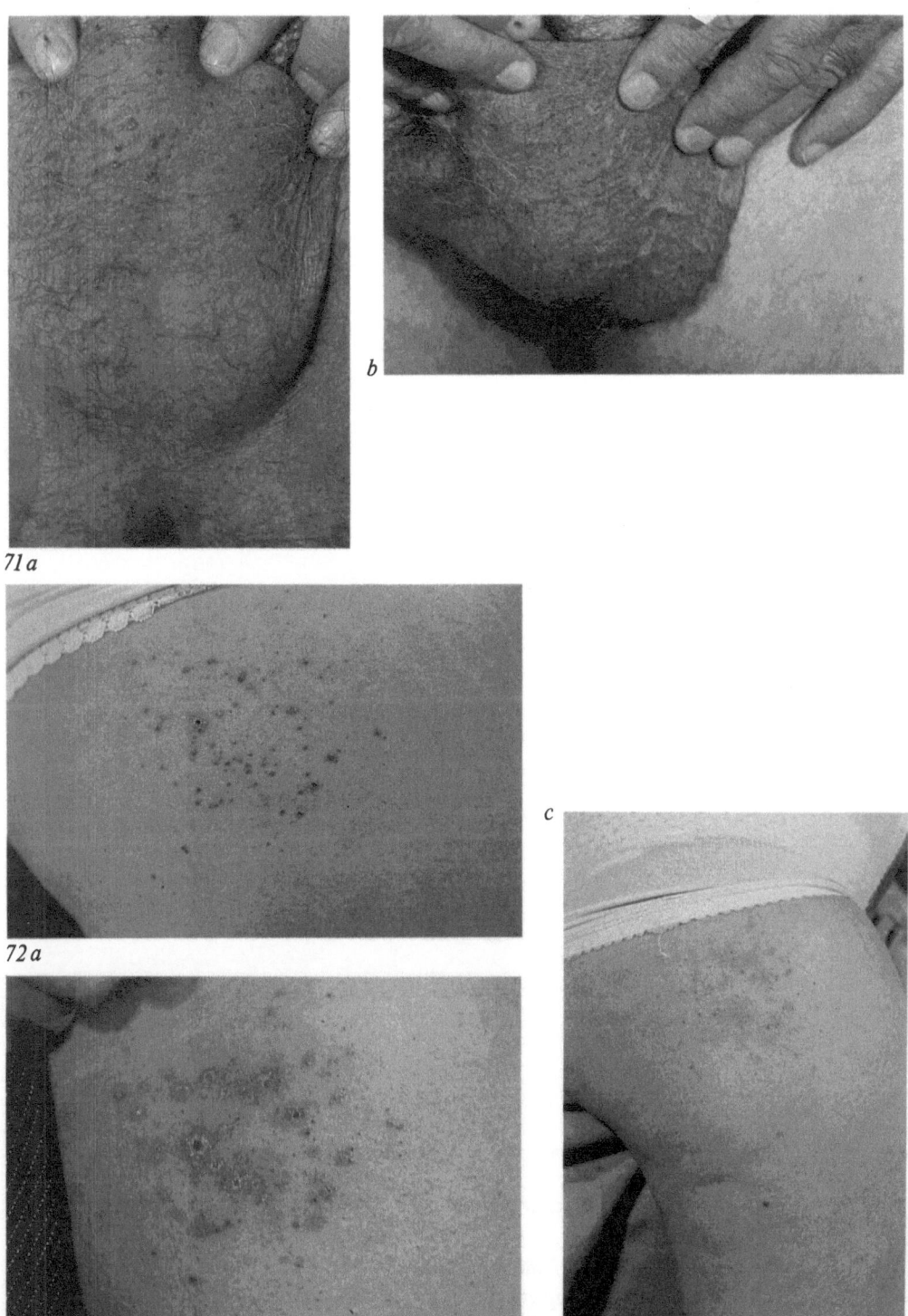

71 a

b

72 a

b

c

2.6 Pyogenic Granuloma

This is usually a solitary papule or nodule that is easily recognized clinically and on occasion may simulate juvenile melanoma in young individuals or Kaposi's sarcoma and amelanotic melanoma in elderly patients.

Figure 73a–d

(a) *Pretreatment view of a vascular lesion clinically diagnosed as a pyogenic granuloma on the right lower lip of an 8-year-old child. Under local anesthesia, the lesion was widely excised and the base was vaporized*

(b) *The histologic section showing a typical pyogenic granuloma*

(c) *The immediate posttreatment view showing a shallow crater on the lower lip*

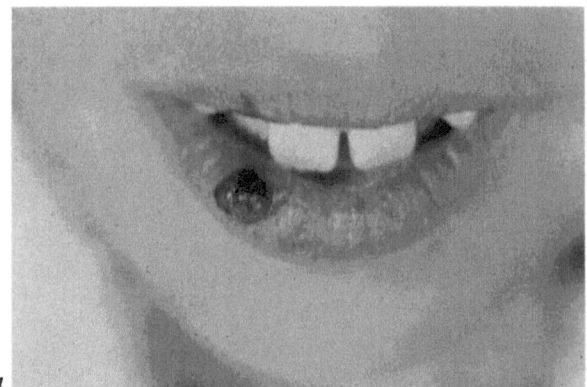

a

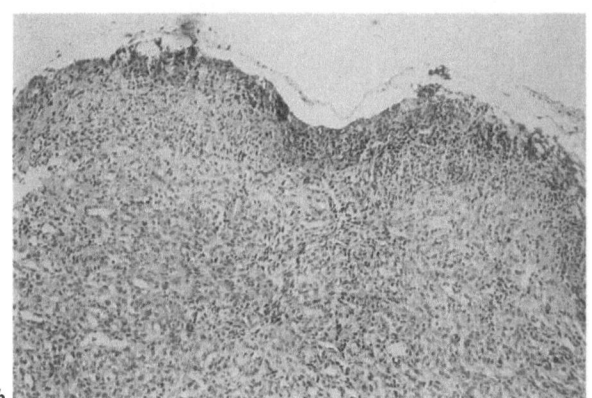

b

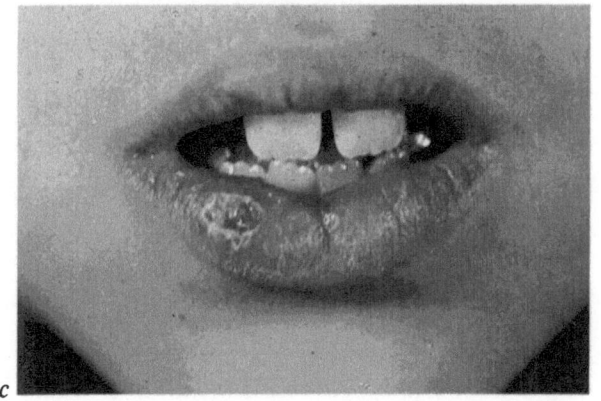

c

Figure 73

(d) Two weeks after treatment, there is complete healing

Figure 74a and b

(a) Pyogenic granuloma of a finger
(b) The clinical appearance 1 month after treatment showing complete healing

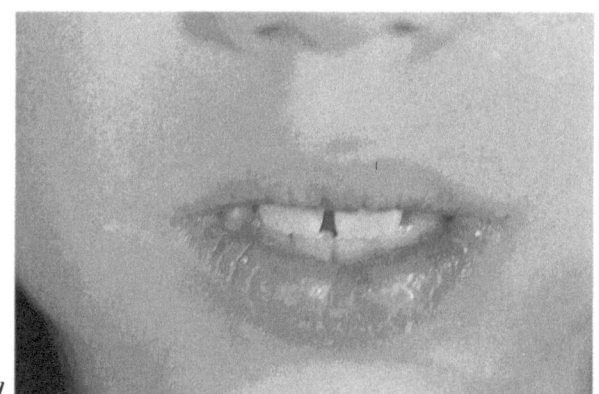

73 d

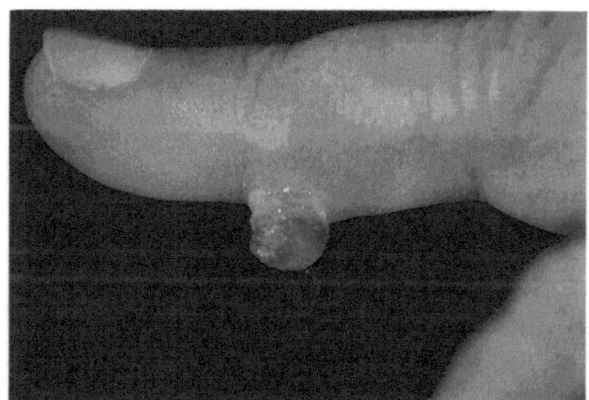

74 a

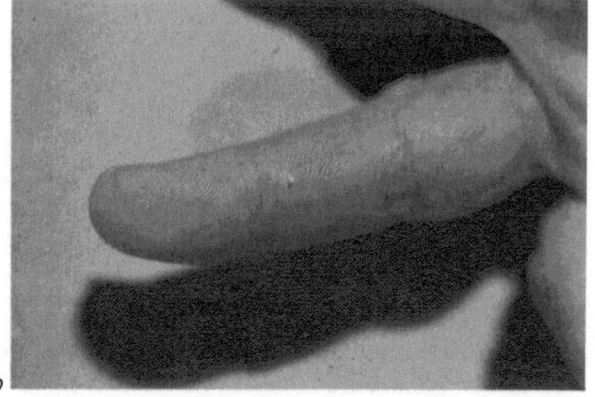

b

2.7 Kaposi's Sarcoma

Figure 75a and b

(a) *Kaposi's sarcoma of the sole of the foot. Because this is a malignant vascular tumor, the CO_2 laser has a dual application in its treatment. Following excision, there was massive bleeding so that a defocused laser beam was necessary to vaporize the crater. Histology showed signs of Kaposi's sarcoma with margins of normal skin free of tumor*

(b) *The clinical view 3 months after treatment. After a follow-up of $1^1/_2$ years there was no recurrence*

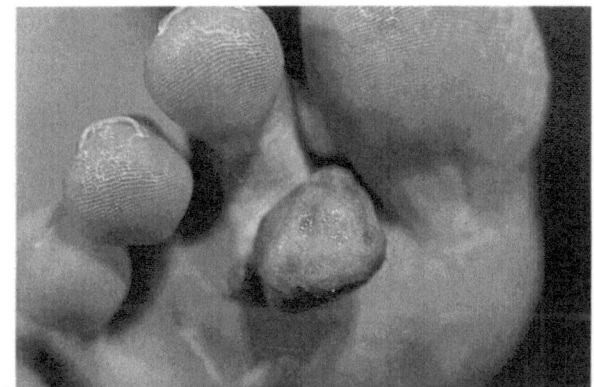

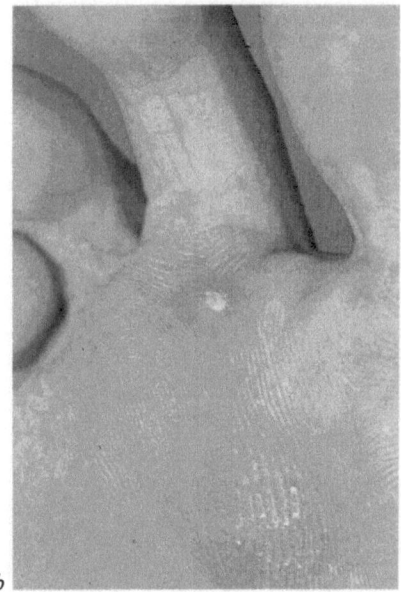

2.8 Benign Cutaneous Lesions

The fact that we are able to avoid suturing with no resultant scarring in these lesions has improved the cosmetic results considerably. When the diagnosis is in doubt, the excised lesions are sent for histologic examination. Multiple lesions can be treated in one session since local anesthesia is not usually required and time is saved by avoiding preparation, suturing, and dressing.

2.9 Seborrheic Keratoses

Seborrheic keratoses are benign skin lesions that occur predominantly in middle-aged and older people. They are usually multiple, sligthly raised, light-brown to deep-black lesions, sharply demarcated, and varying in size from tiny to large tumors, which rarely disappear spontaneously. The therapeutic modalities for the treatment of seborrheic keratoses are simple curettage, electrodessication, cryosurgery, or surgical excision under local or general anesthesia. All these methods are effective to some degree but are impaired by several disadvantages, namely, incomplete removal of the lesions, need for repeated procedures, inability to control the precise depth of destruction of the tissues, excessive postoperative pain, and occasional unsatisfactory scarring. The many advantages of the CO_2 laser make this modality superior to others in the treatment of solitary as well as multiple seborrheic keratoses. With the use of the CO_2 laser, the lesions are vaporized with a continuous defocused beam (when local anesthesia is required). For small multiple lesions, no local anesthesia is required and a pulsed beam is used.

Figure 76 a–c

(a) Seborrheic keratosis of lower abdomen in a female patient
(b) The histologic section showing a typical seborrheic keratosis
(c) One month after treatment. The scar is still red and soft. It may take another month for the redness to disappear completely

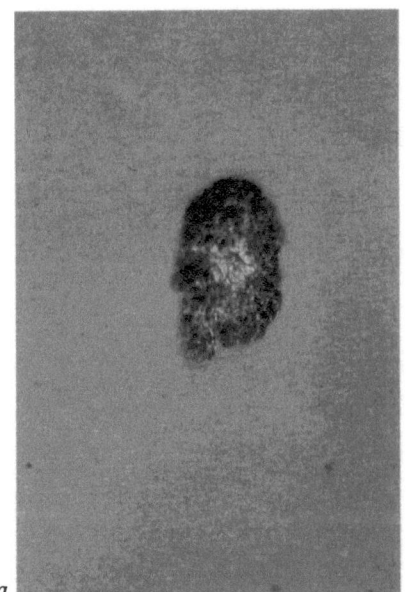

a

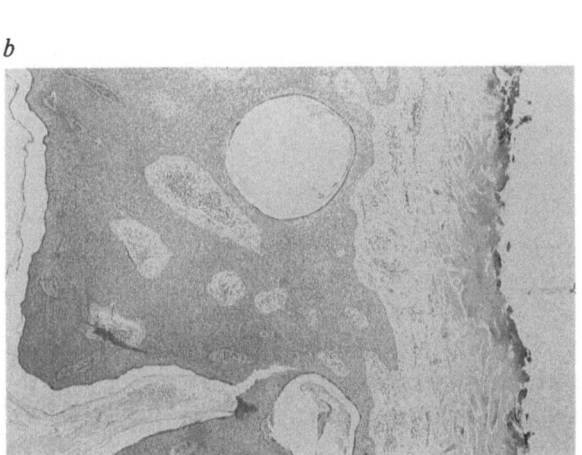

b

c

Figure 77a and b

(a) *Seborrheic keratosis of the cheek, verified by histologic examination*
(b) *The appearance 3 weeks later*

Figure 78a and b

(a) *Multiple seborrheic keratoses of the lower abdomen and pubis in a 36-year-old patient. The lesions were excised in a single session under local anesthesia*
(b) *The clinical appearance 2 months after treatment. Note slightly hypopigmented scars*

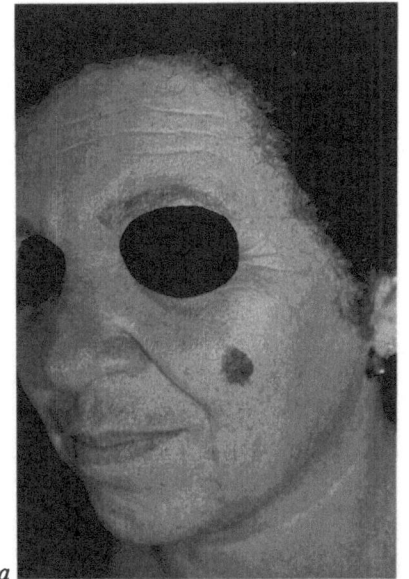
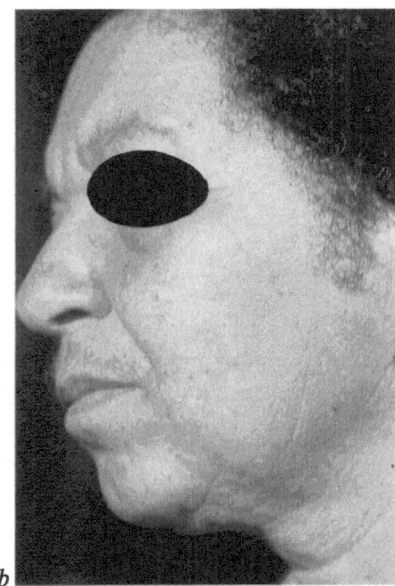

77 a *b*

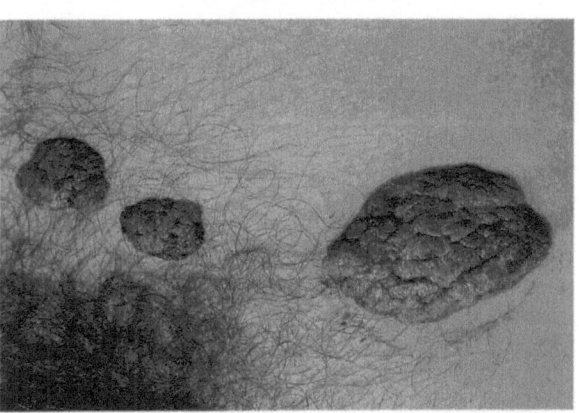

78 a

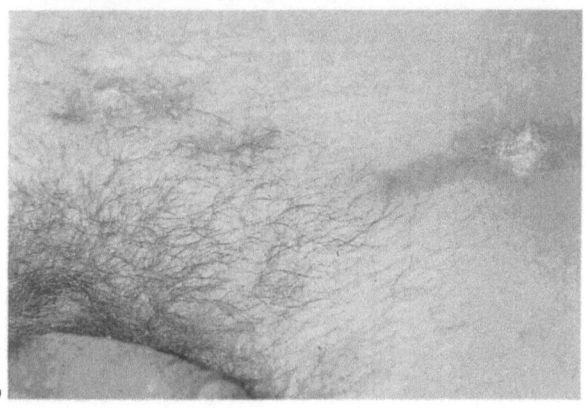

b

Figure 79 a–e

(a) Multiple seborrheic keratoses of the back in a 56-year-old female patient. In this patient, because of the large number of lesions general anesthesia was used. More than 400 lesions were vaporized in the first session

(b) The condition 1 month after treatment

(c) The condition 2 months after treatment, some lesions still existed so a second treatment was carried out under local anesthesia

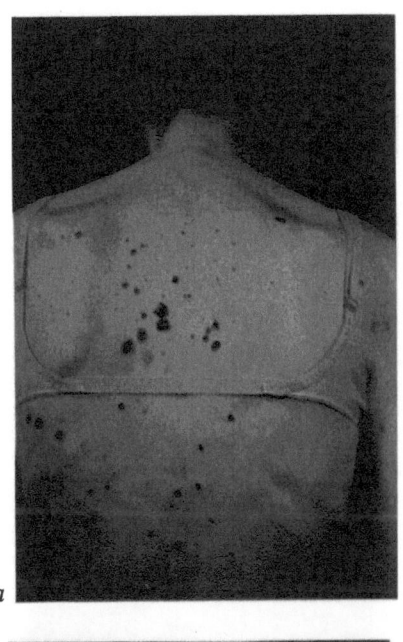

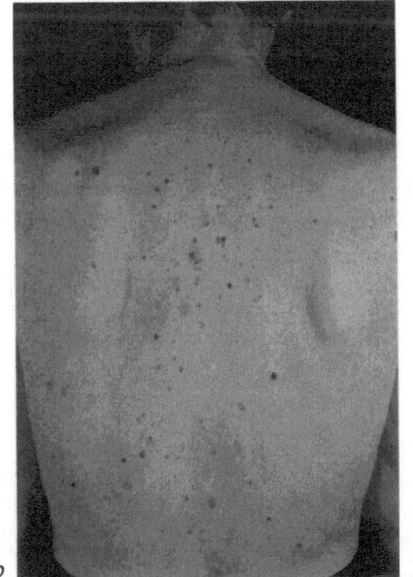

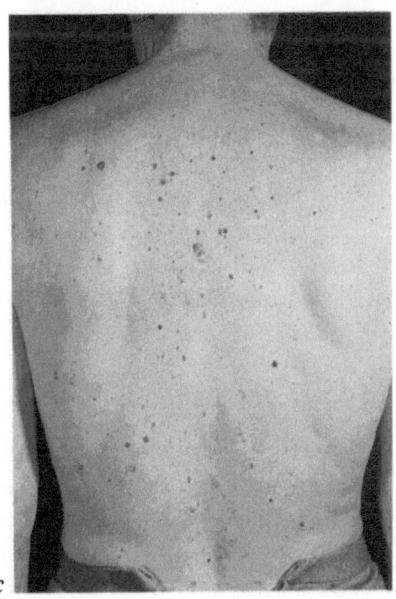

Figure 79

(d) *The condition 2 weeks after the second treatment. Three lesions on the middle of the back were excised with a scalpel and sent for histologic examination. In this figure, we can compare the healing processes after using laser and scalpel. Note that the crust still remains on the scalpel wounds*

(e) *Two months after the second treatment. There are still several hypopigmented scars*

Figure 80 a–c

(a) *Multiple seborrheic keratoses of the chest*

(b) *The condition 24 h after treatment. Multiple craters can be seen covered with exudation and scab formation*

(c) *The condition 6 months after treatment. Note that there is no hypertrophic scarring on the sternal region as compared with an old hypertrophic laparotomy scar*

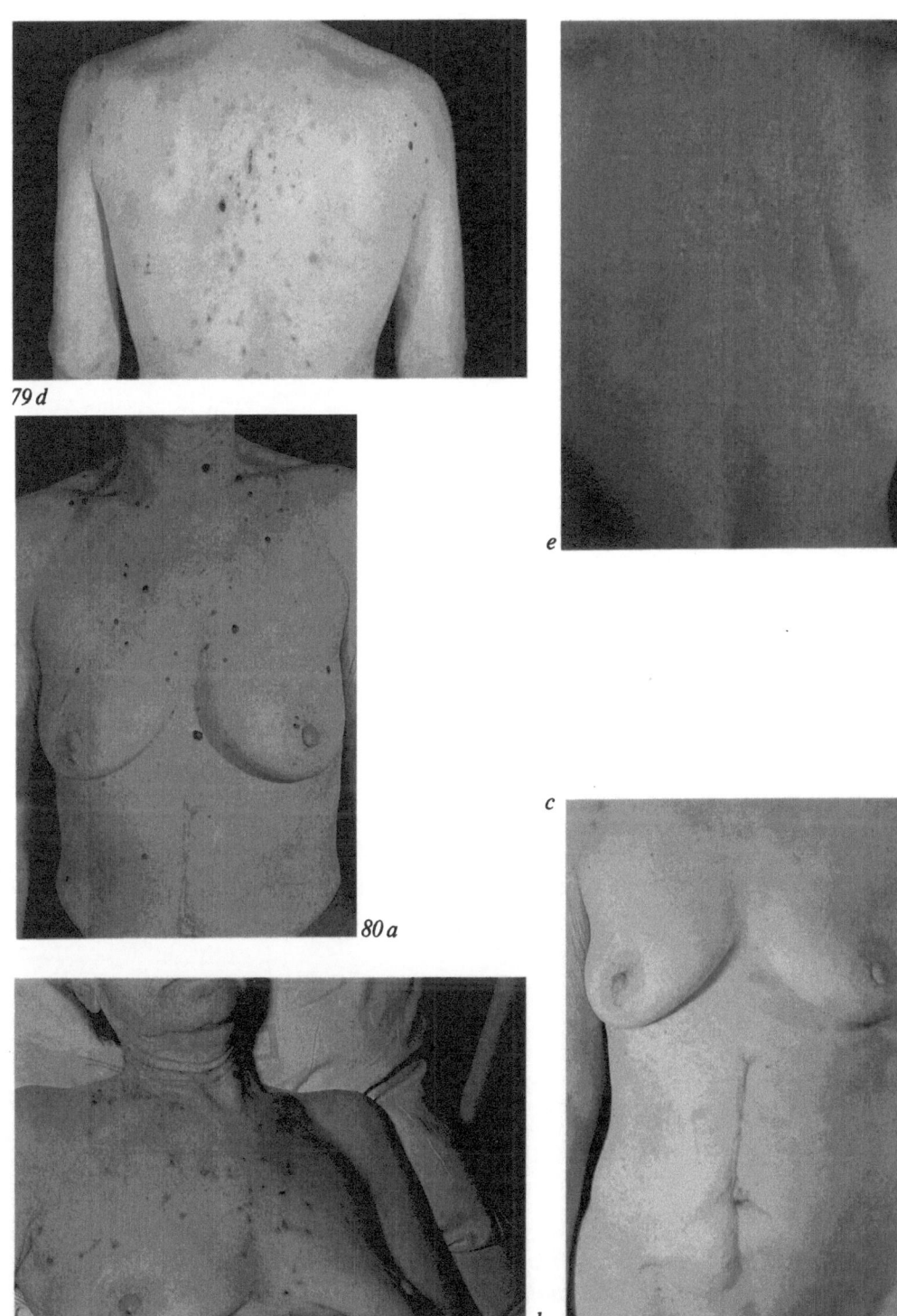

79 d

80 a

e

b

c

2.10 Papillomatosis

Figure 81a and b

(a) Pretreatment view of papillomatosis of the face in an elderly woman. In such cases, the lesion can simply be vaporized with a defocused beam

(b) The condition 2 weeks after treatment

2.11 Keratoacanthoma

Although this is a benign lesion that is usually resolved spontaneously within several months in a typical case, it must be adequately excised because of the possibility of confusion with other malignant lesions.

Figure 82a and b

(a) A large keratoacanthoma of the cheek in an elderly woman. The lesion was widely excised and histologic examination confirmed the clinical diagnosis

(b) The condition 3 months after treatment

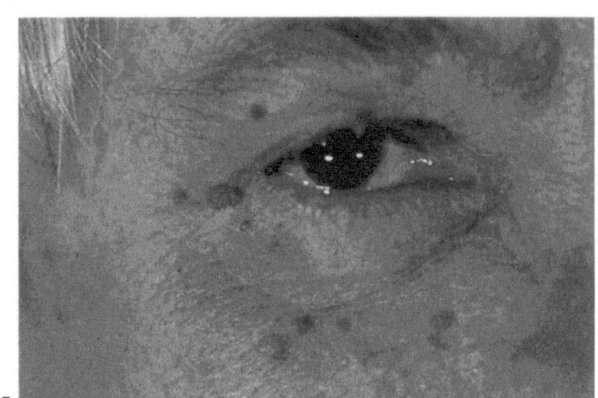

81 a

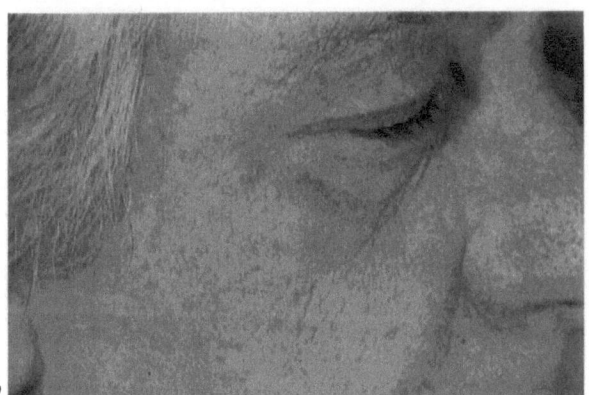

b

82 a

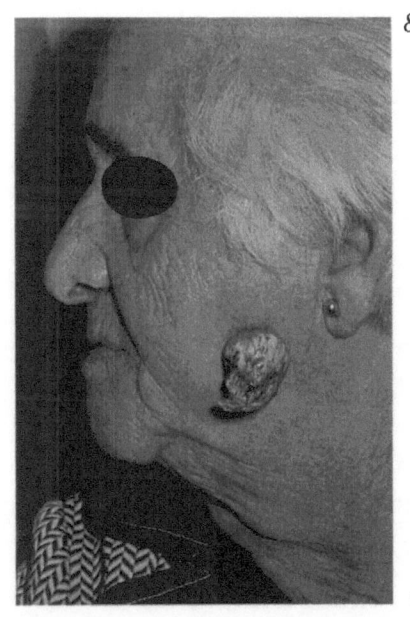

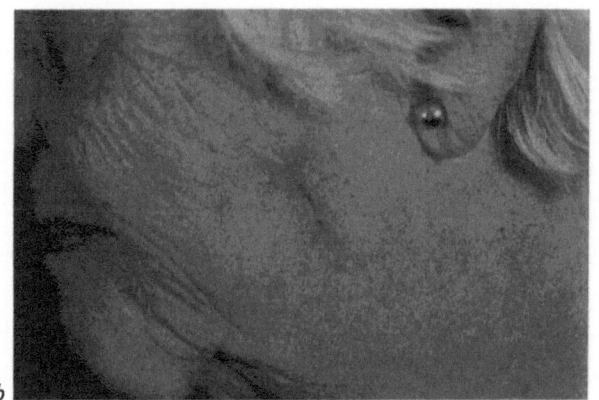

b

2.12 Skin Appendages

The introduction of the CO_2 laser in the treatment of small accessory auricles, skin tags, and accessory digitis has enabled us to undertake their removal in the immediate postnatal period without anesthesia or bleeding. The treatment is very rapid and no ligatures are required. The wounds heal without obvious scarring.

Figure 83a and b

(a) Accessory auricle in a newborn
(b) The appearance 4 months after treatment

Figure 84a and b

(a) Accessory digitis in a 40-year-old woman
(b) The condition 1 month after treatment

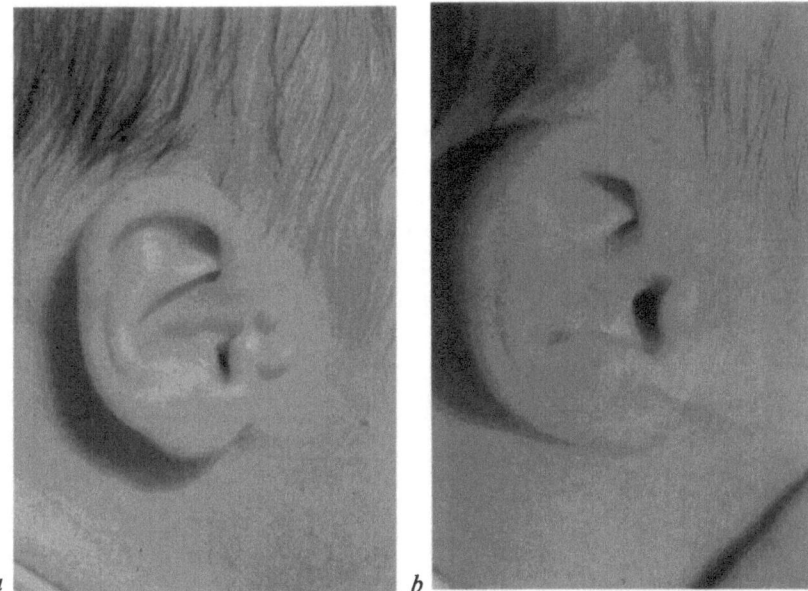

83 a
b

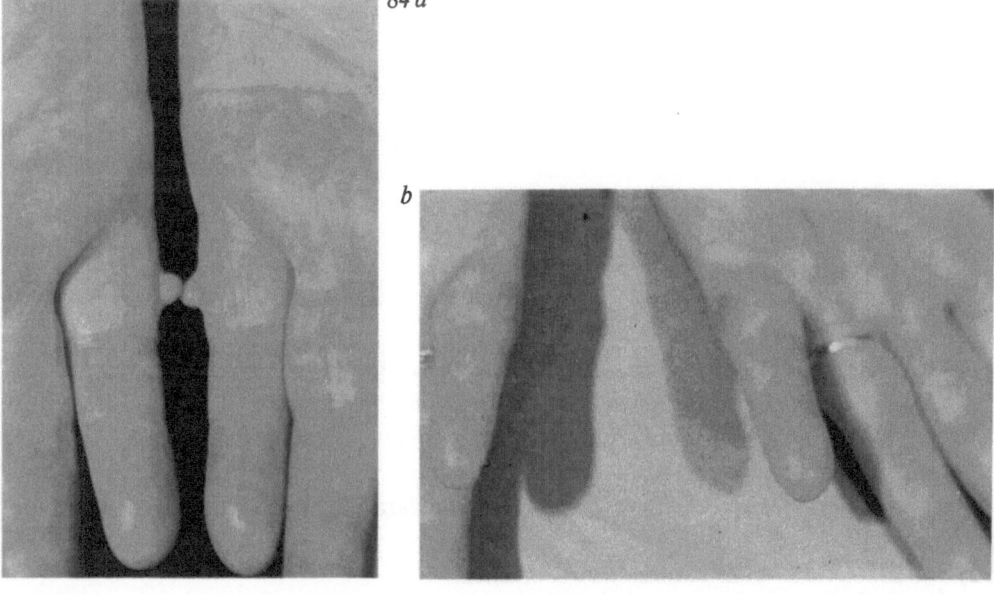

84 a
b

2.13 Verrucae

Verrucae are very common viral infections of the skin. Although the vast majority of warts will undergo spontaneous resolution in the course of time in most cases, the treatment is for cosmetic reasons. Most of the patients referred to us for laser treatment had already undergone various attempts at removal by application of different methods of treatment. The treatment of these lesions with the CO_2 laser is very rapid and results in a significant decrease in recurrence rate. The cosmetic results are also superior. In the rare event of recurrence, a second procedure results in complete cure.

Figure 85 a–c

(a) *Verrucae vulgaris of the fingers. Because verrucae are epidermal structures, only superficial vaporization is performed as deeply as necessary. Care is taken not to damage the dermis. Following the treatment, there is no need to limit the patient's activity (movement, washing, etc.)*

(b) *Two months later, there is a complete healing of the skin with good texture and color*

(c) *Follow-up of 1 year showed no recurrence*

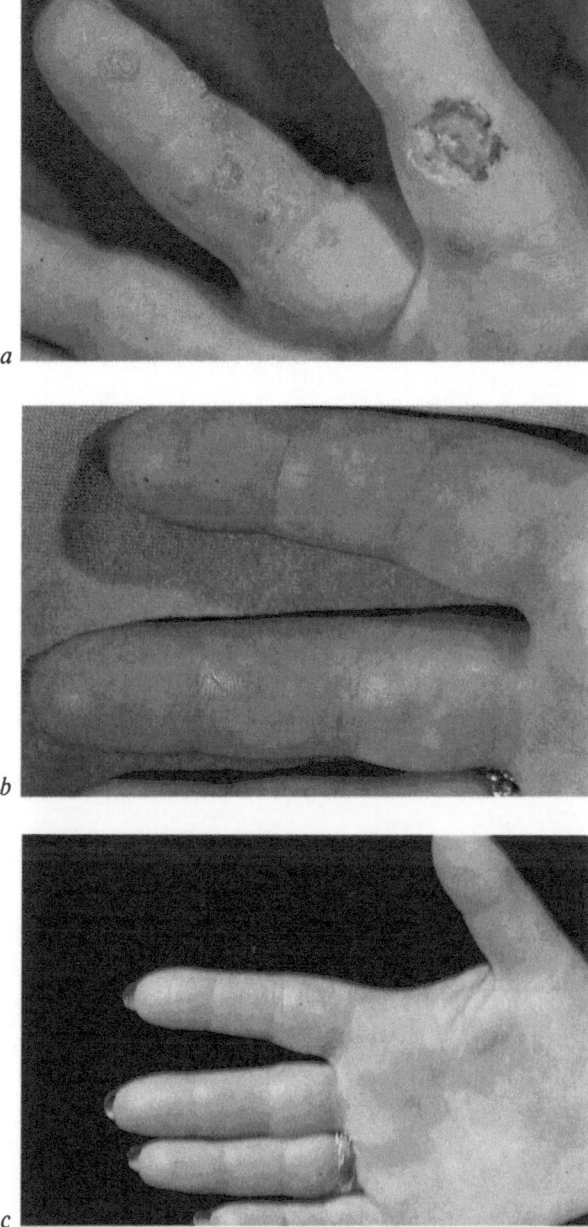

Figure 86a and b

(a) *Diffuse verrucae of the lower face in a 10-year-old child. In this case, general anesthesia was performed because of the age of the patient*

(b) *The condition 1 month later*

Figure 87a–c

(a) *Multiple warts on the thumb of a 12-year-old girl. All the lesions were vaporized under local anesthesia*

(b) *The appearance 1 month later*

(c) *The appearance 1 year later*

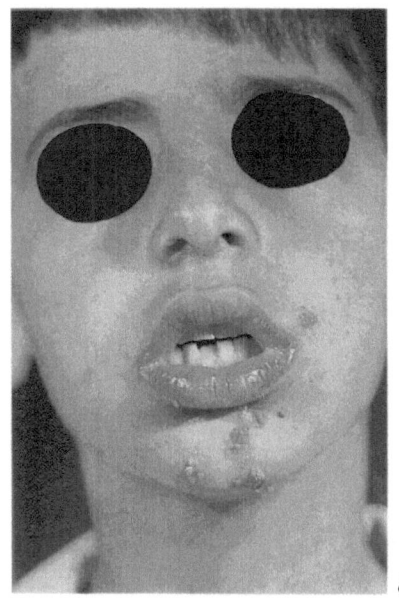

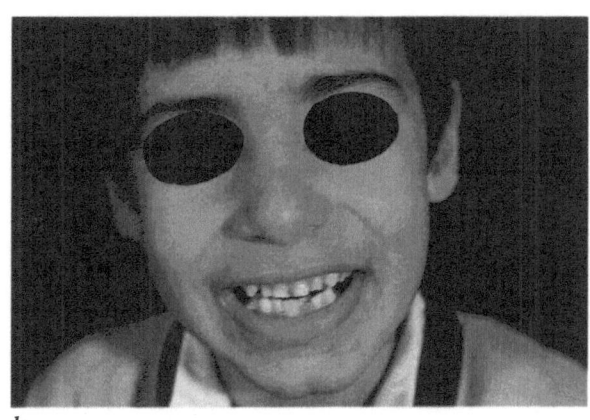

b

86 a

87 a

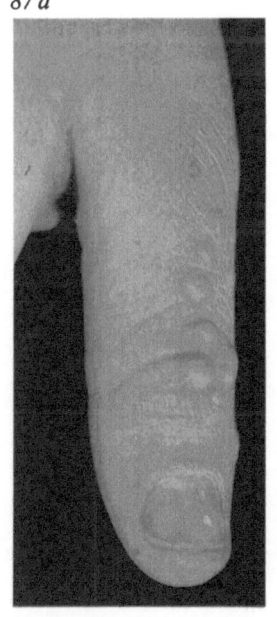

b

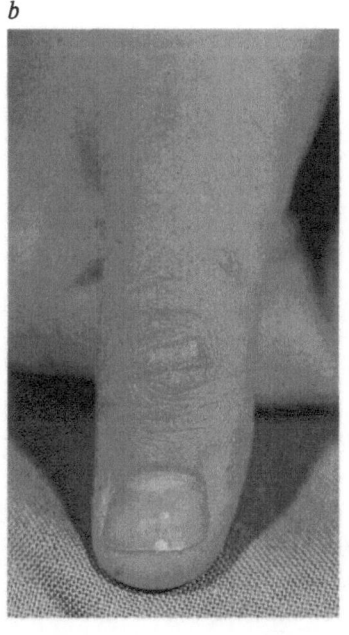

c

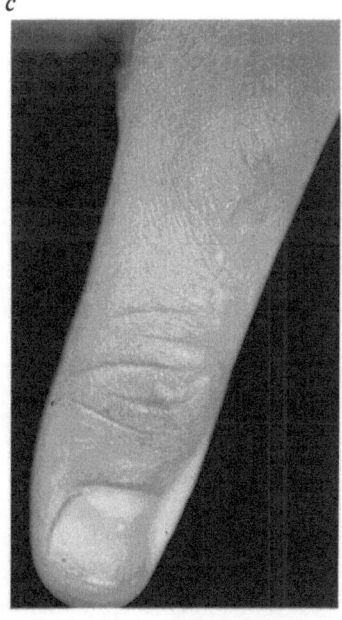

2.14 Intradermal Nevi

Figure 88 a–c

(a) Intradermal nevus of the nostril
(b) The appearance immediately after excision
(c) The appearance 1 month later showing a good cosmetic result

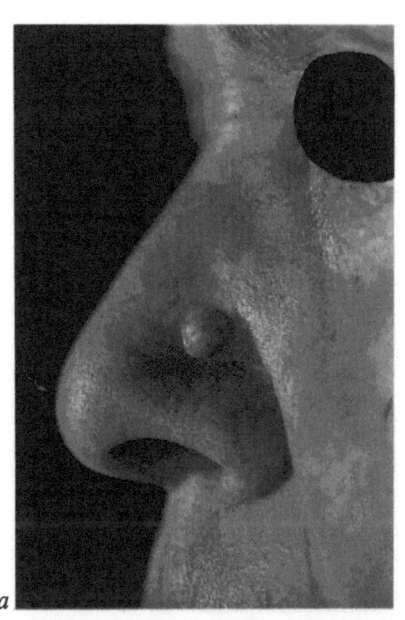

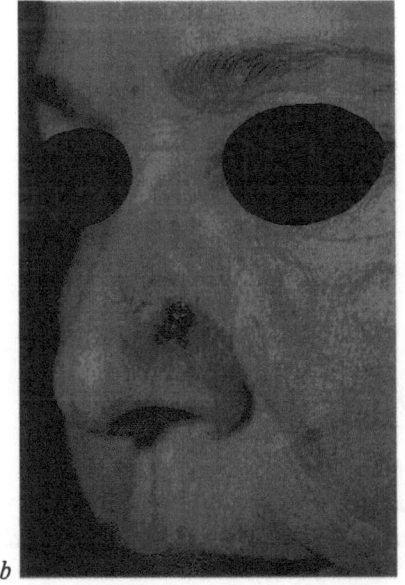

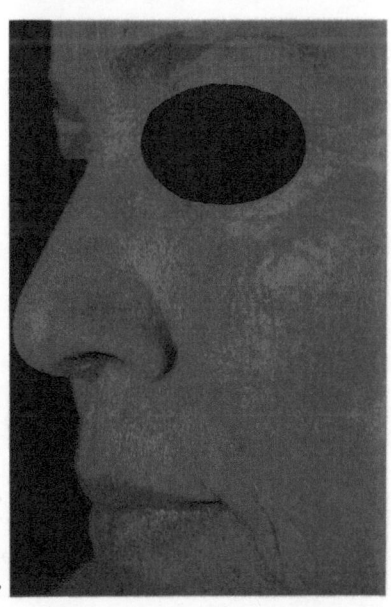

Figure 89 a–d

(a) Large intradermal nevus on the eyebrow region. Excision and vaporization of the base were performed
(b) The histologic section showing intradermal nevus
(c) The appearance 1 month after treatment, a pink scar still exists
(d) The appearance 1 year after treatment. Note that there is no deformity of the eyebrow

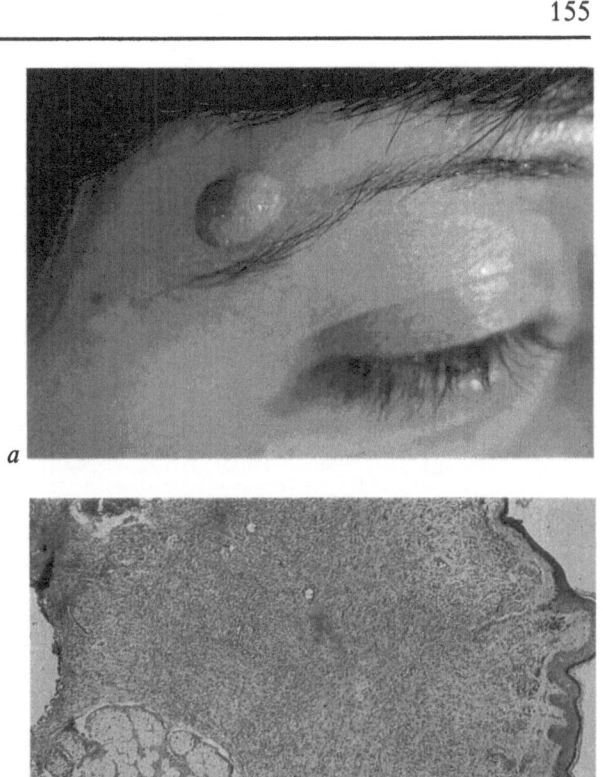

a

b

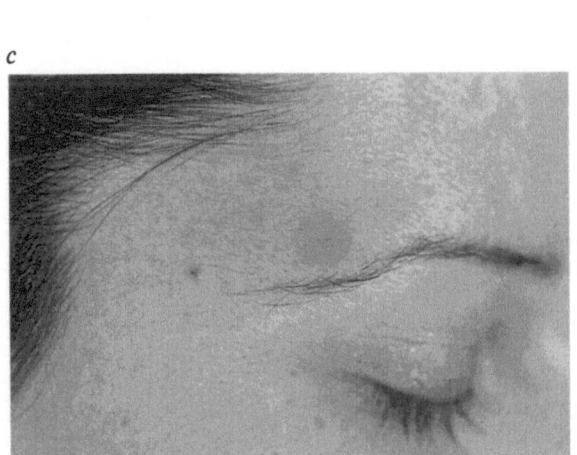

c

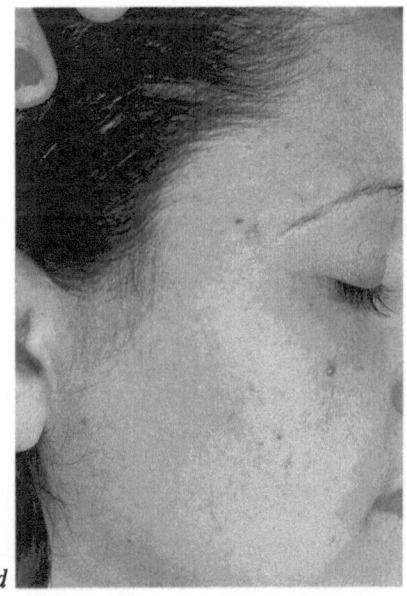

d

Figure 90 a–c

(a) Multiple intradermal nevi on the face
(b) The appearance immediately after treatment
(c) The appearance 1 month later

Figure 91 a and b

(a) Intradermal nevus of scalp
(b) The appearance 1 month later

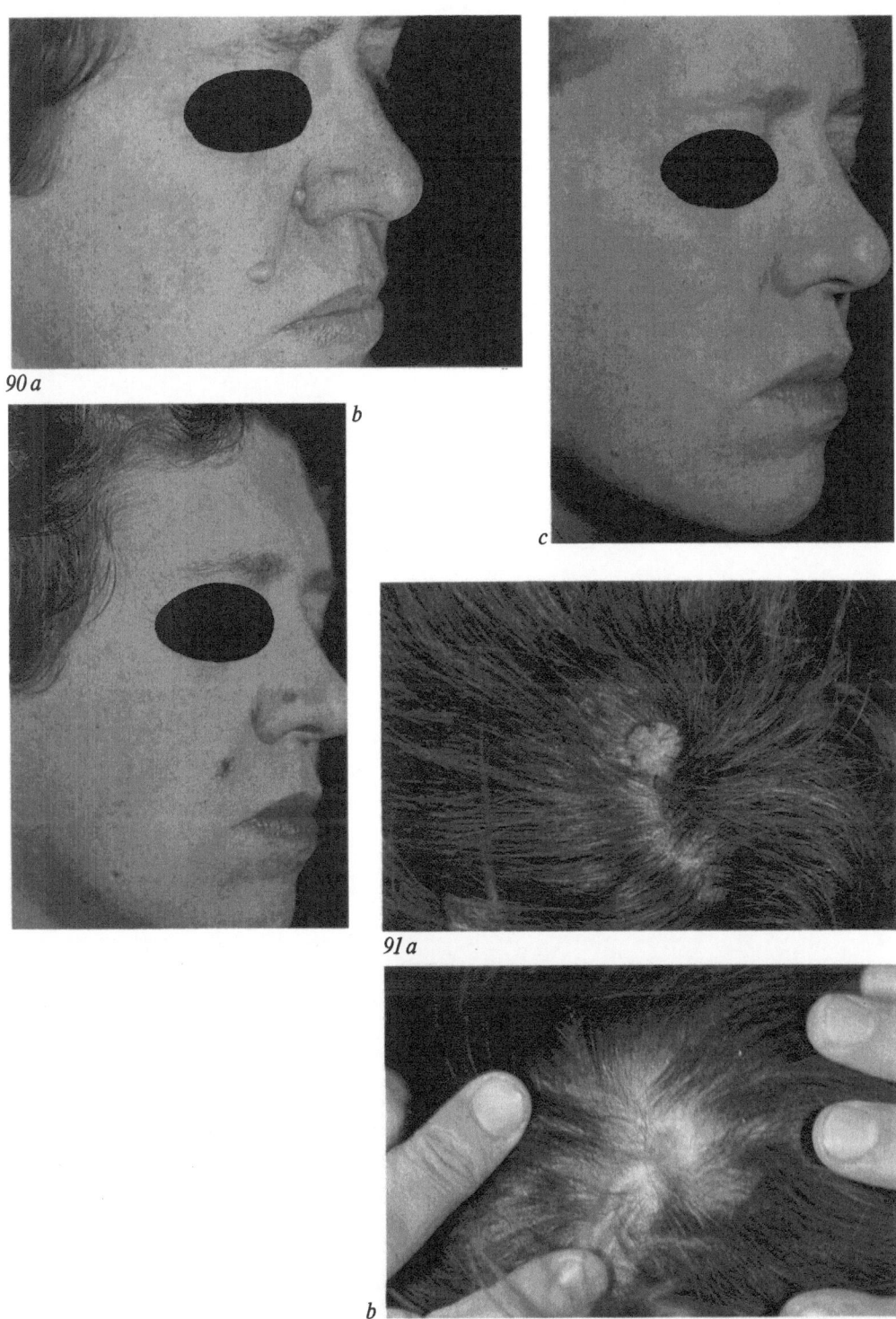

90 a

b

c

91 a

b

2.15 Pigmented Nevi

Figure 92a–d

(a) Papillomatous nevus of the eyebrow region in a young woman. Note that conventional excision in this area would have resulted in a deformity of the eyebrow

(b) The appearance immediately after excision and vaporization. Note absence of bleeding

(c) The appearance 2 weeks after treatment. A small crater still exists

(d) The appearance 1 month after treatment

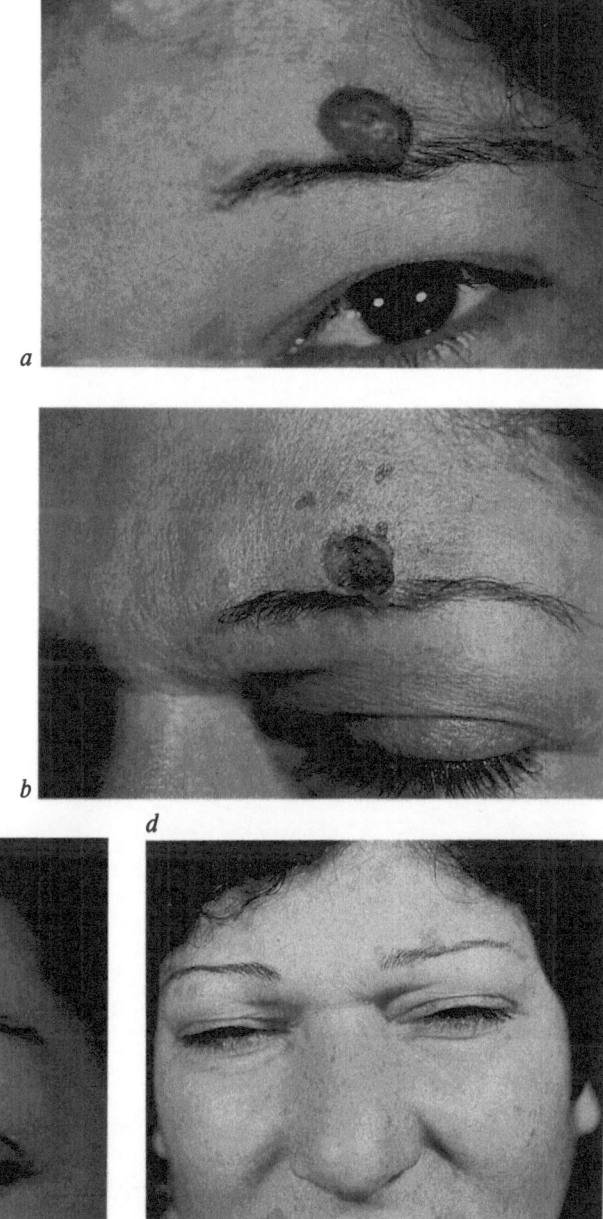

Figure 93 a–c

(a) Papillomatous nevi of the lip
(b) The appearance immediately after treatment
(c) The appearance 1 month later

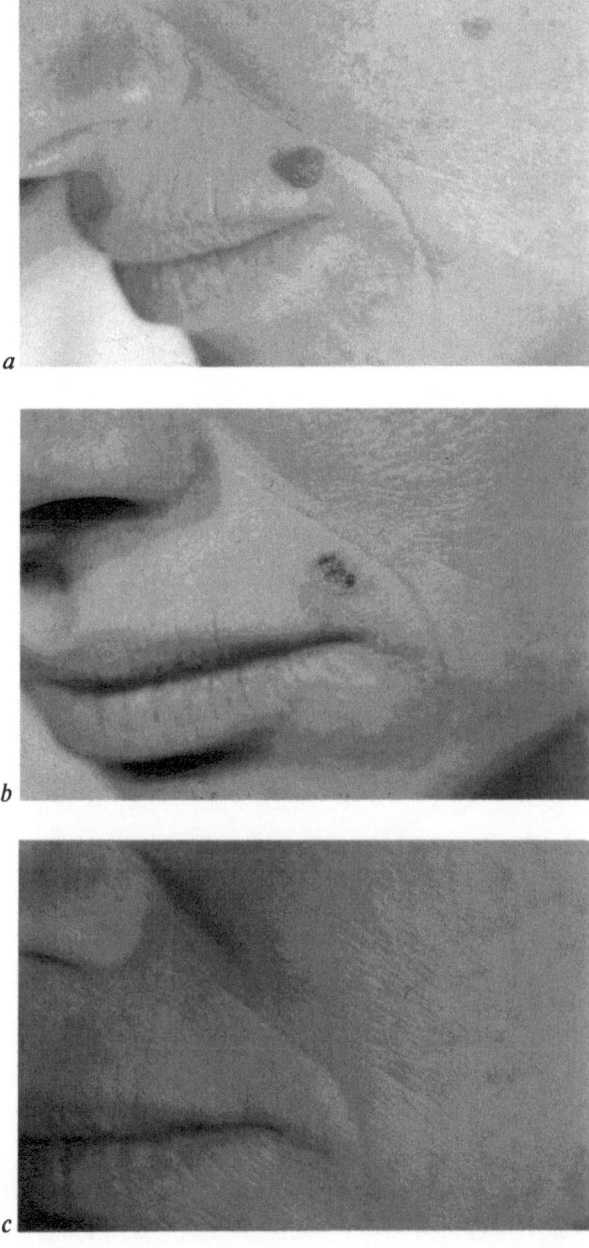

Figure 94a–c

(a) Papillomatous nevus of the face
(b) The appearance immediately after excision and vaporization
(c) The appearance 1 month later (close-up)

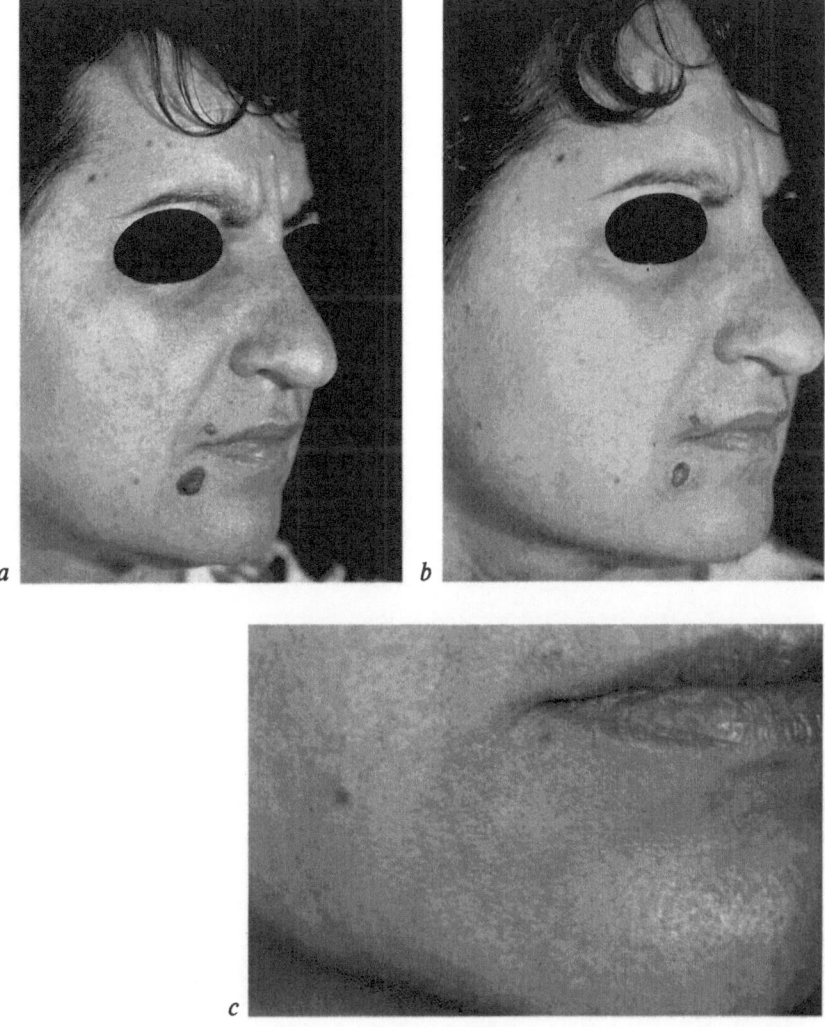

a

b

c

Figure 95a and b

(a) Papillomatous nevi on the back of the neck
(b) The appearance 1 month after treatment

Figure 96a and b

(a) Multiple papillomatous nevi of the back
(b) The appearance 1 month later. Only slight hypopigmentation still exists

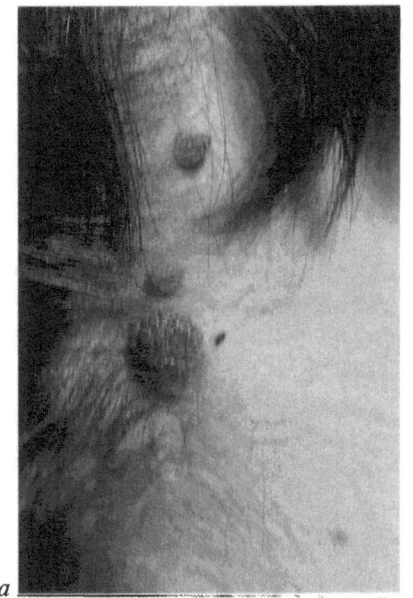

95 a

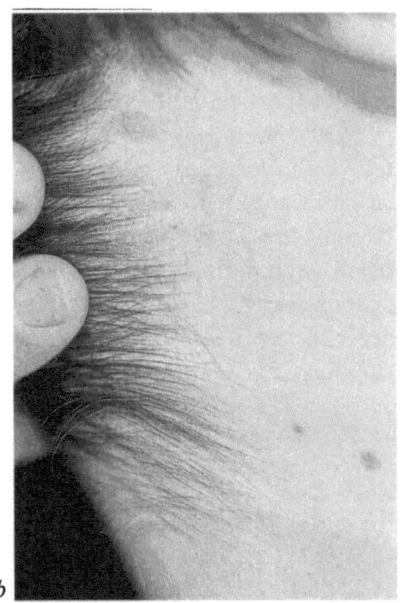

b

96 a

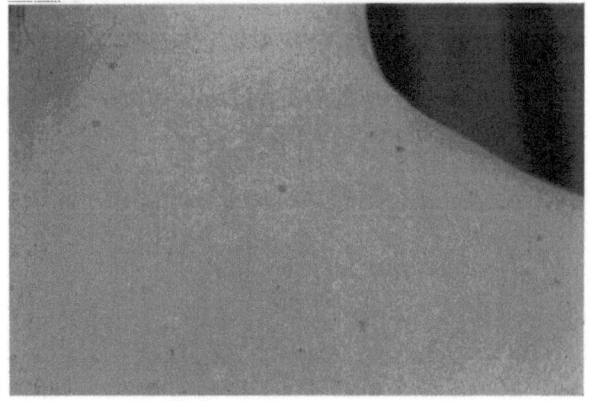

b

2.16 Basal Cell Carcinoma

Small superficial and well-defined basal cell carcinomas lend themselves to satisfactory treatment with the CO_2 laser. This, in our opinion, is superior to either electrocautery or cryosurgery. After taking a biopsy, the lesions are vaporized or excised. In cases where recurrences occur, a second procedure is found to be satisfactory. It must be borne in mind, however, that this procedure should be confined to small, well-defined, unilocular, and superficial lesions. For large or poorly defined lesions, excisional surgery is recommended also using the CO_2 laser but sutures are necessary to close the surgical defect. In all cases, histologic confirmation of the diagnosis is required

Figure 97a and b

(a) A deep basal cell carcinoma of the cheek in a female patient suffering from "farmer's skin." As the disease was multilocular in this patient, it was decided to attempt to treat the deeper lesions as well as the superficial lesions with the CO_2 laser. It should be noted that the patient had undergone multiple procedures in the past

(b) The appearance 1 year later

Figure 98a–d

(a) Adenocystic basal cell carcinomata of the upper eyelid and the canthal area

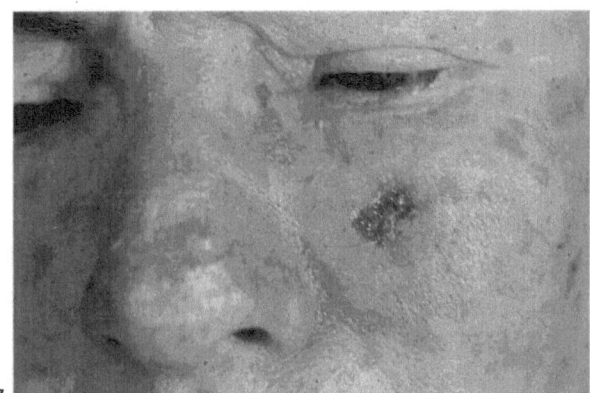

97 a

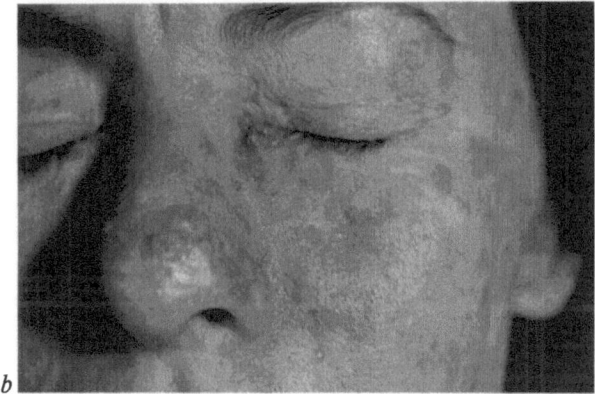

b

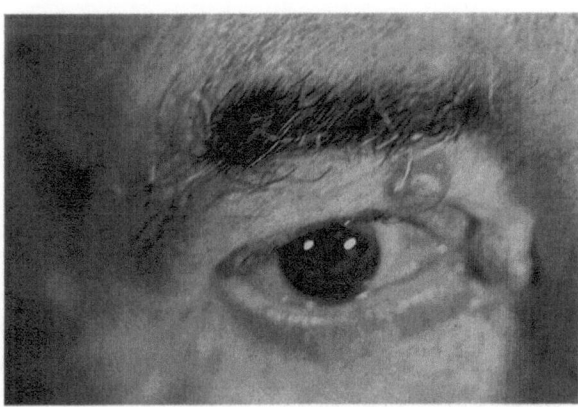

98 a

Figure 98

(b) The histologic section of one lesion confirming the clinical diagnosis

(c) The appearance immediately after excision followed by vaporization of the bed

(d) The appearance 1 year later; note no recurrence, good cosmetic results, and no eyelid deformity

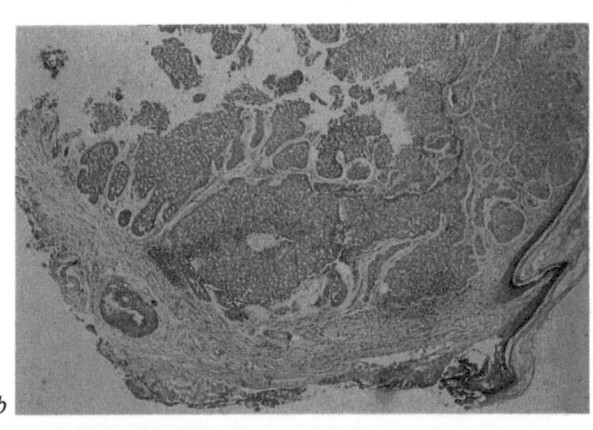

b

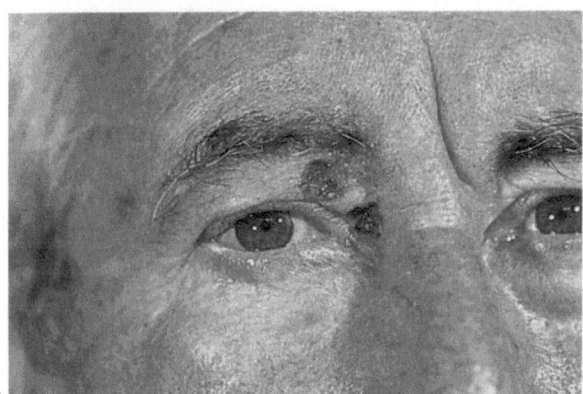

c

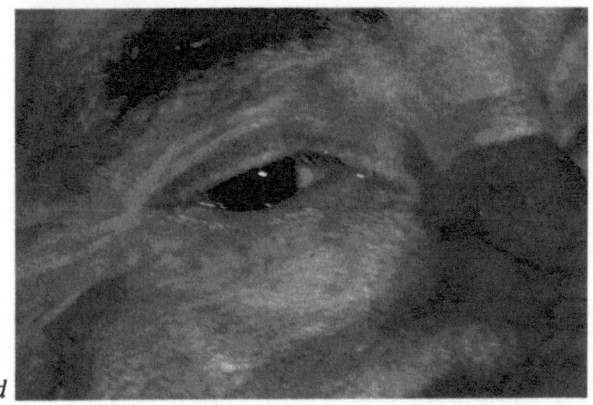

d

Figure 99 a–e

(a) A typical rodent ulcer of the nostril
(b) The histologic section showing basal cell carcinoma
(c) The appearance immediately after treatment

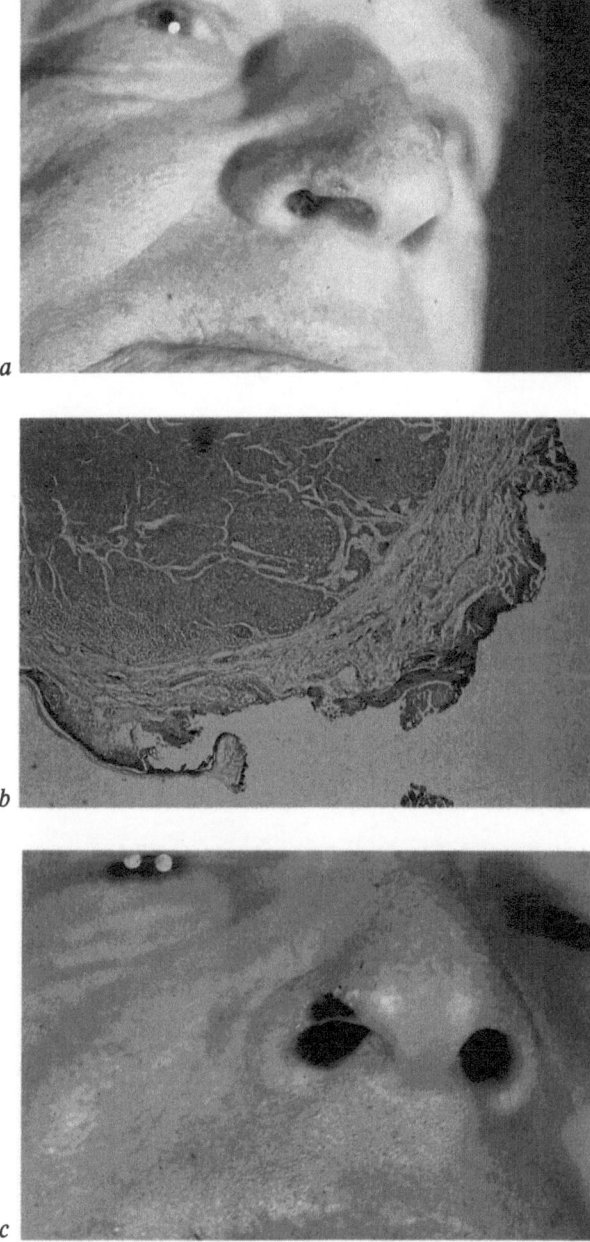

a

b

c

Figure 99

(d) The appearance 2 weeks after treatment
*(e) The appearance 4 months after treatment. Follow-up of 2 years revealed no recur-
rent disease*

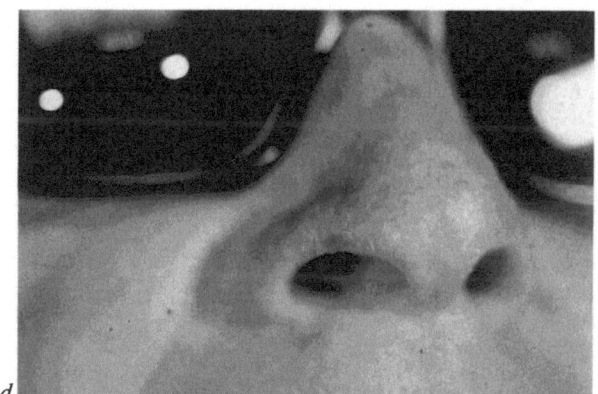

d

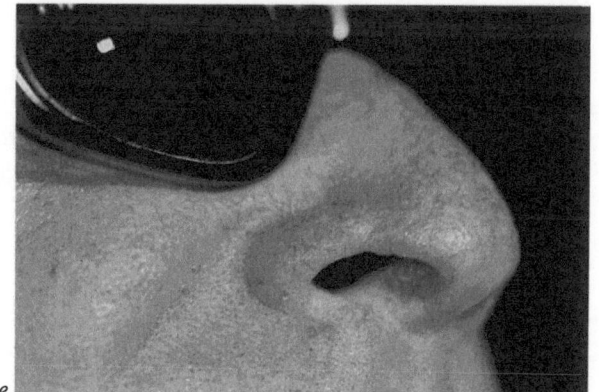

e

Figure 100a and b

(a) *An elderly patient with basal cell carcinoma of the cheek and diffuse actinic changes. Superficial vaporization of the actinic changes was performed and the basal cell carcinoma was excised*
(b) *The appearance 3 months later*

Figure 101a and b

(a) *Adenocystic basal cell carcinoma of the upper lip*
(b) *The appearance 1 month after treatment*

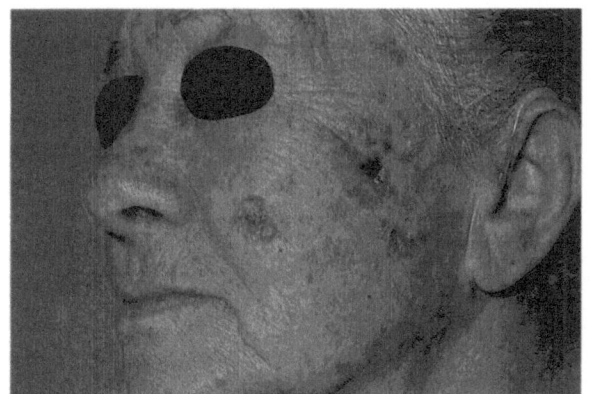

100 a

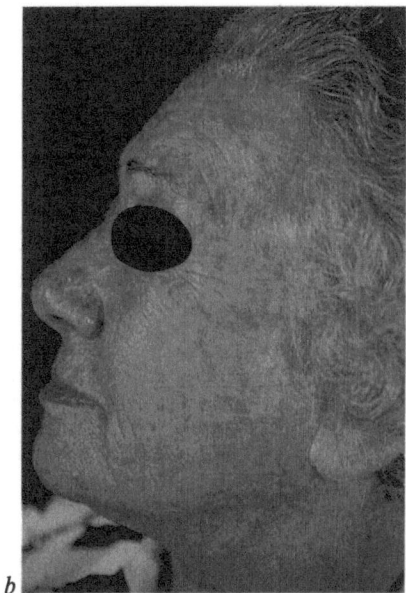

b

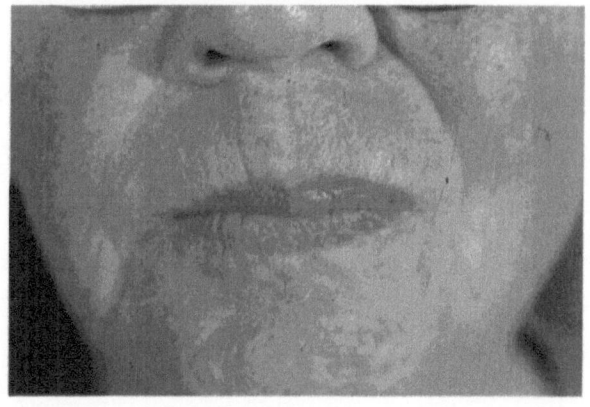

101 a

b

2.17 Juvenile Melanoma

Figure 102a and b

(a) A case of juvenile melanoma of the arm clinically diagnosed as hemangioma in a 2-year-old child

(b) The appearance 4 months after excision and suture. The importance of obtaining histologic confirmation of the clinical diagnosis should be stressed

Figure 103a and b

(a) A case of juvenile melanoma of the auricle clinically diagnosed as a pyogenic granuloma

(b) The condition 1 month after excision and vaporization

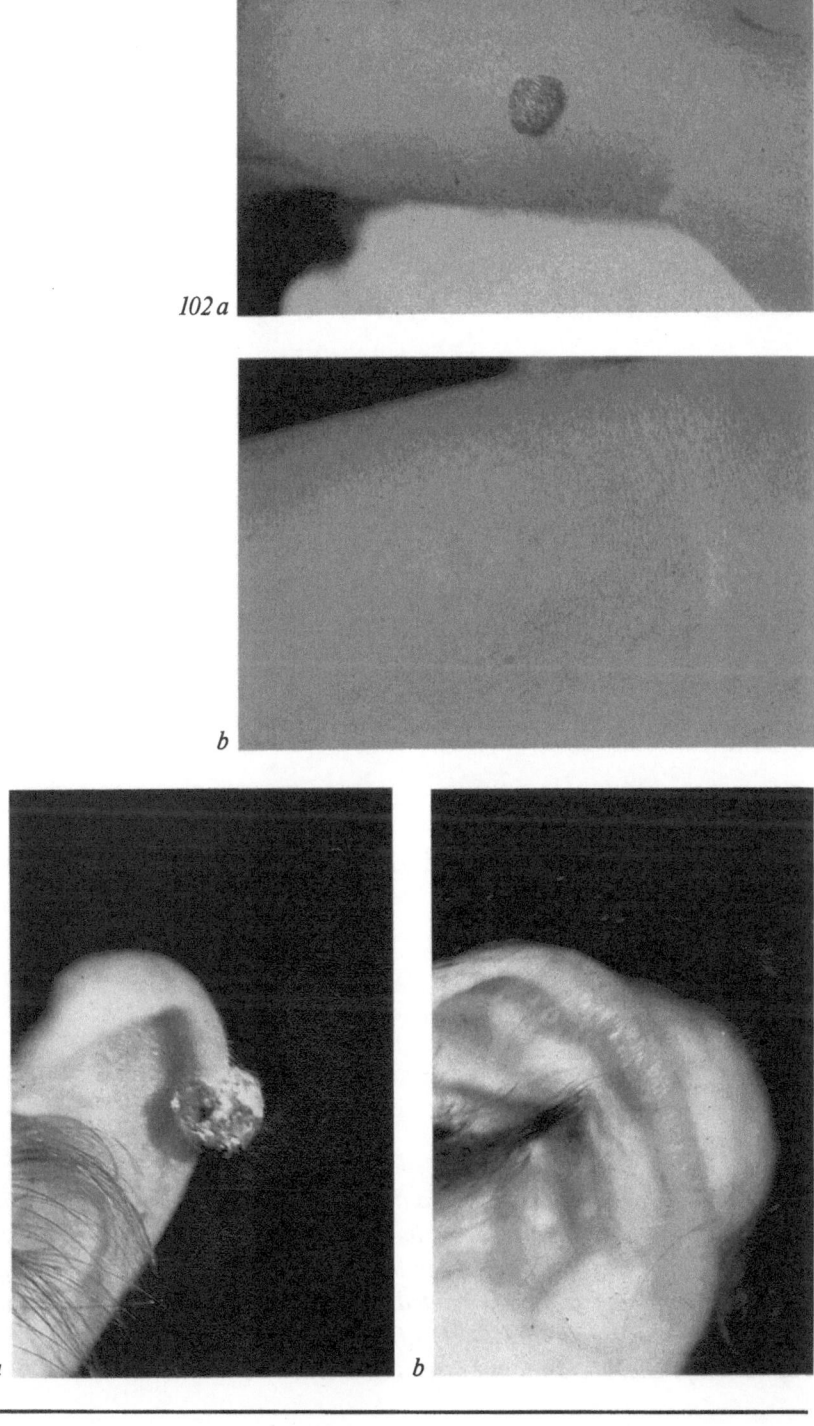

102 a

b

103 a

b

2.18 Xanthelasma

Figure 104 a–c

(a) A middle-aged woman with xanthelasma on the upper eyelid near the inner canthus. When the laser beam is used in such cases, the eyes are covered with wet sponges. The lesion was vaporized using a defocused beam. When larger plaques are present, laser surgical excision is needed with primary suturing to repair the surgical defect

(b) The appearance immediately after treatment

(c) The appearance 1 month after treatment

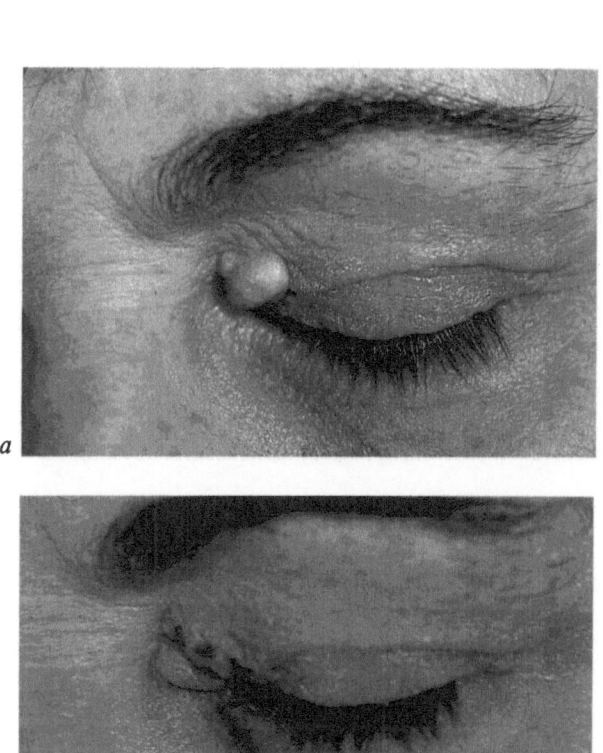

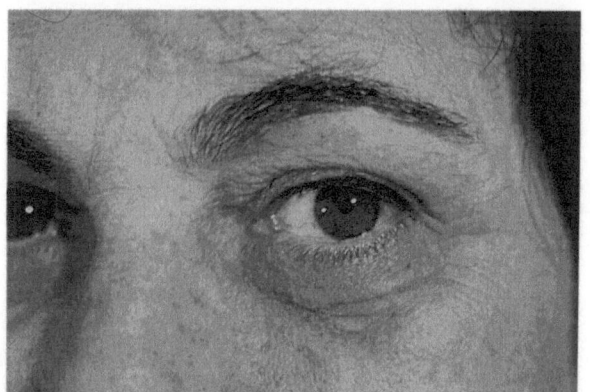

2.19 The CO_2 Laser in Oral Surgery

The vascularity of the tongue and buccal mucosa is well known and operations in these areas are often accompanied by copious bleeding followed by hematomas and edema. The ability of the CO_2 laser to seal blood vessels while cutting results in a reduction of bleeding during surgery. It is also known that after CO_2 laser surgery, there is less postoperative pain and edema and a reduced tendency toward scar formation. These advantages make the CO_2 laser a superior modality in the treatment of lesions of the tongue and oral mucosa.

Figure 105 a–c

(a) Fibroma on the tip of the tongue. The lesion was excised without anesthesia
(b) The immediate posttreatment view. There is no bleeding or pain
(c) Two weeks after the treatment there is complete healing

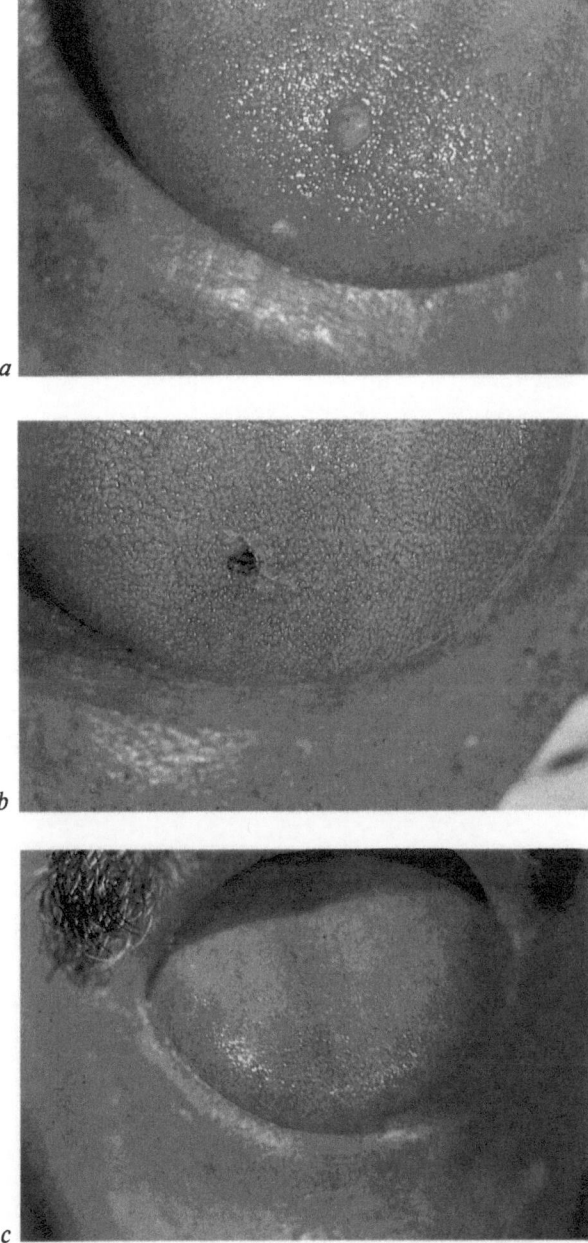

Figure 106 a and b

(a) Fibroma of the buccal mucosa
(b) The clinical view, 1 month after treatment

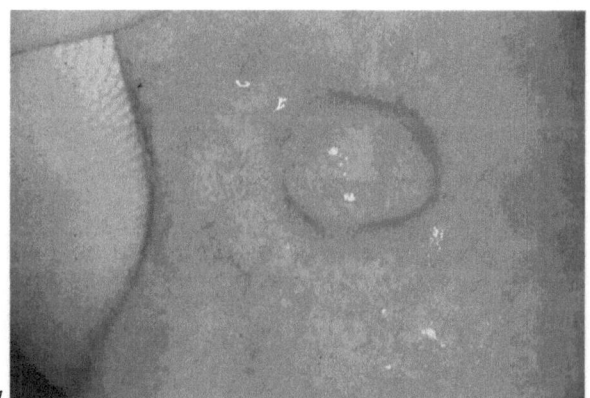

a

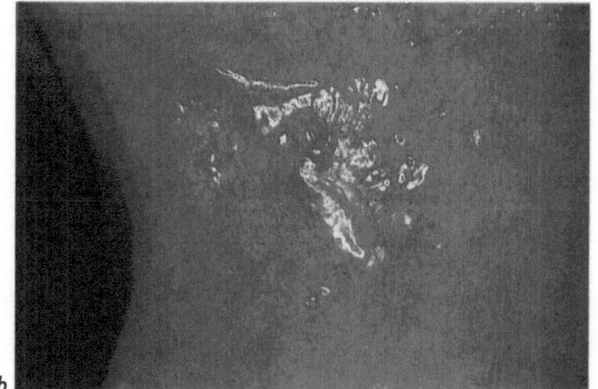

b

Figure 107a and b

(a) *A huge benign tumor at the base of the tongue. The tumor was excised under local anesthesia*

(b) *The view 1 month after treatment. Although the lesion was very large there was no need to suture the wound because of the sealing effect of the laser beam*

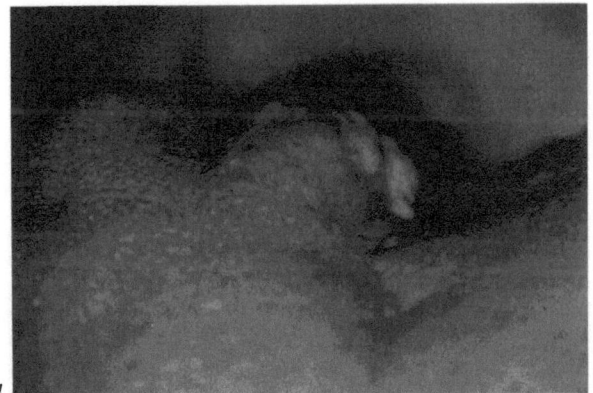

a

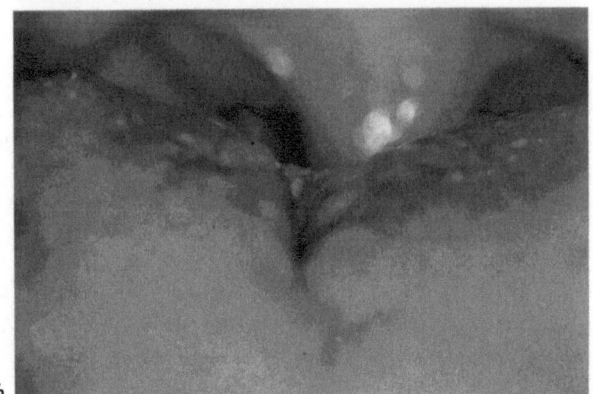

b

2.20 Neurofibromatosis

Superficial neurofibromas are dermal tumors varying in size and shape that may appear anywhere on the body and are merely a cosmetic deformity. When these lesions occur diffusely, surgical removal is greatly facilitated by laser vaporization of the small lesions and laser excision and vaporization of the larger ones. It is worthy of note that it is important in this procedure to remove the neurofibromata entirely in order to avoid recurrences.

Figure 108a and b

(a) Diffuse neurofibromatosis of the back in a young woman
(b) The appearance 2 months later

Figure 109a–c

(a) Diffuse neurofibromatosis of the chest. In this case, the procedure was performed under general anesthesia because of the large number of lesions
(b) The appearance 24 h after treatment
(c) The appearance months after treatment

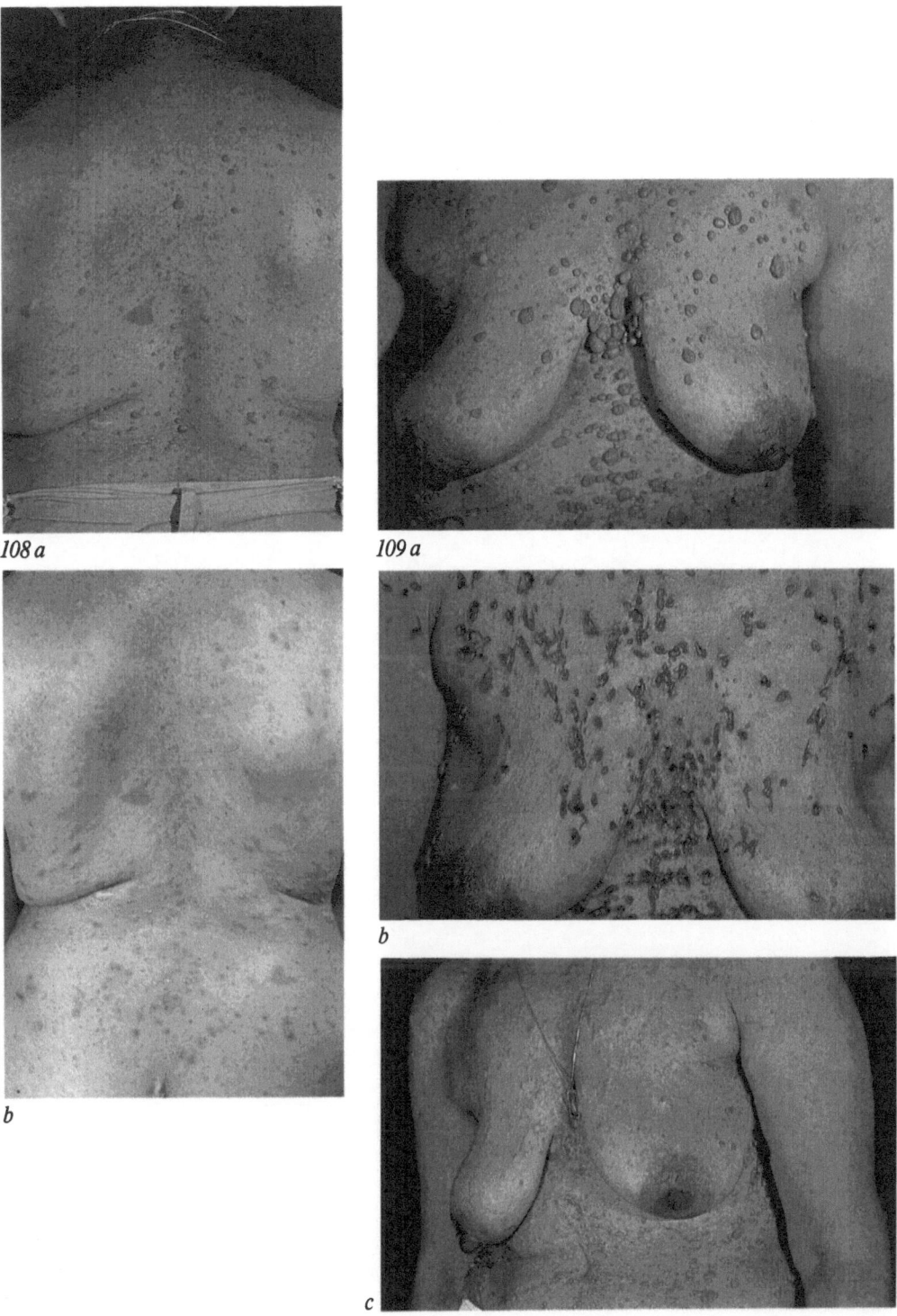

108 a

109 a

b

b

c

2.21 Leukoplakia

Figure 110 a and b

(a) Leukoplakia of the lower lip. Superficial vaporization is performed in such cases
(b) The appearance 1 month later

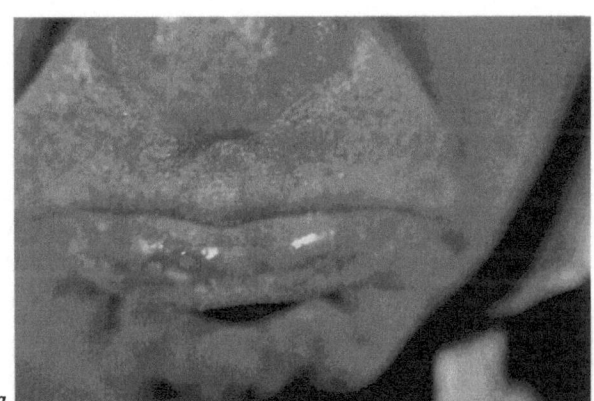

a

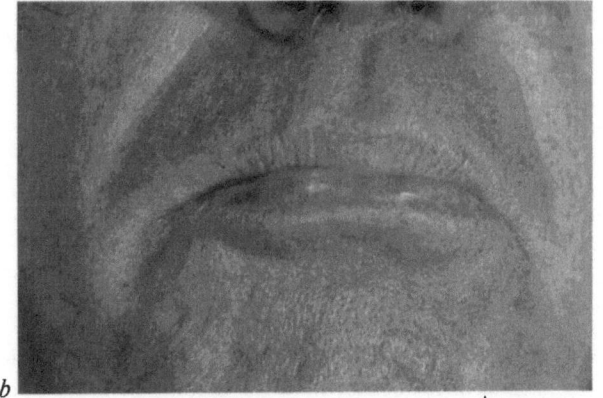

b

2.22 Computerized Laser Dermabrasion

It has long been felt that the treatment of extensive superficial lesions could be enhanced by the application of a CO_2 laser scanner. Accordingly, a computerized laser scanner was developed for this purpose and is used in our department for the dermabrasive treatment of large superficial lesions, such as – port-wine stains and other extensive hemangiomas requiring vaporization, xeroderma pigmentosum, farmers' skin, Bowen's disease, postacne and burn scars, extensive pigmentation (giant nevi, Becker's nevi, nevus unius lateris, and decorative tattoos), large seborrheic keratoses, leukoplakia, pruritus vulvae and and ani, crural and decubitus ulcers, and burn eschars. The system consists of a microprocessor control unit and an electric motor-driven scanner.

Method of Treatment. The surgeon can demarcate the area to be treated using the He-Ne laser guide light controlled by a joystick. The shape of the area to undergo dermabrasion is then stored in the memory. The laser is then set to work within the required limits and only the depth of the lesion requires further control. By depressing the laser "on" footswitch, the CO_2 laser beam is activated automatically, within the predefined boundaries in two directions, with preselected power level, speed, spot size, and line density. The debris is then wiped away with a wet pad and the depth of abrasion estimated. This can be repeated until the required depth is reached.

This procedure enables us to perform dermabrasion, with an accurate control of depth and extent, without bleeding, and without the scattering of debris and blood, more precisely and evenly than with conventional methods. The healing process is rapid with minimal pain and good cosmetic results.

Figure 111

The computerized laser scanner

Figure 112a–m

(a) – (c) A patient with xeroderma pigmentosum. Under general anesthesia, computerized laser dermabrasion was performed

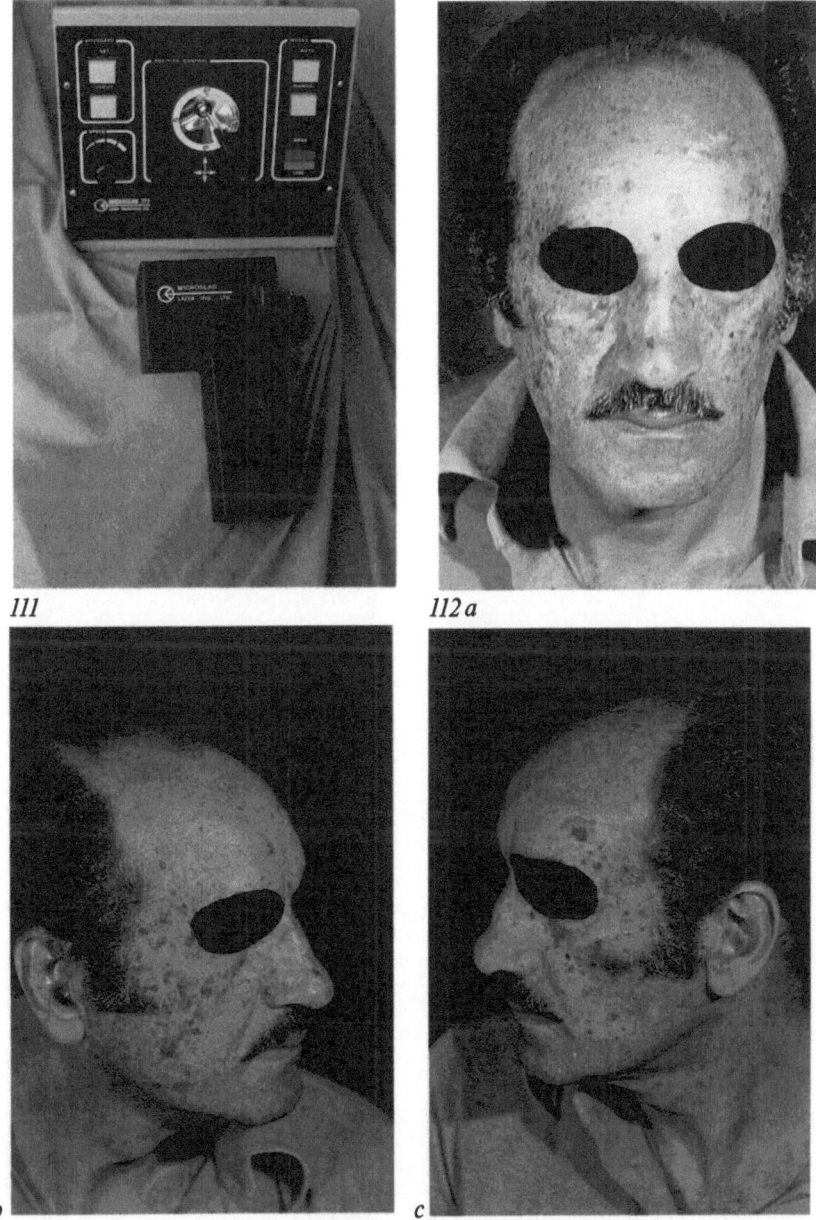

111

112a

b

c

Figure 112

(d) — (f) The appearance, 48 h after surgery

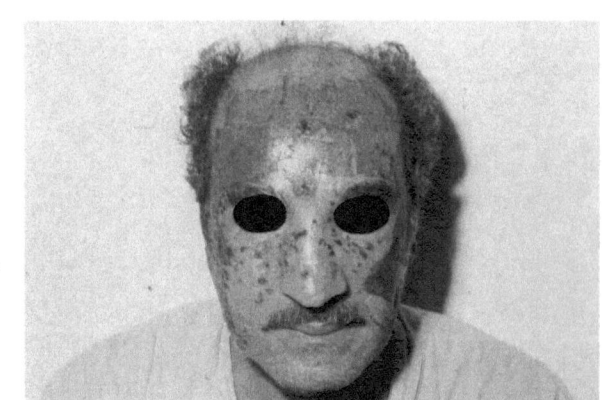

d

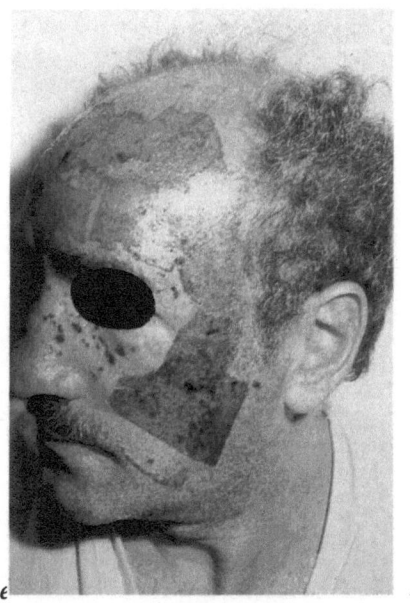

e

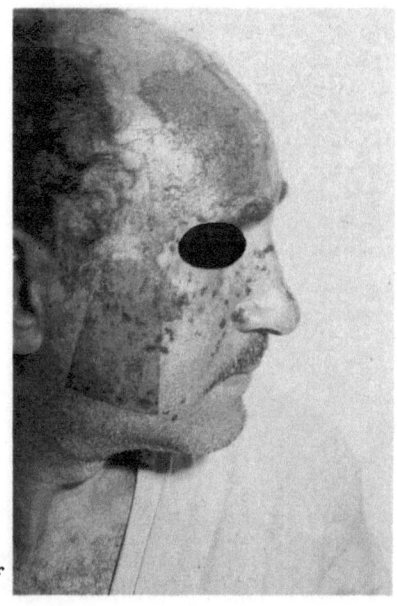

f

Figure 112

(g) — (i)   The appearance, 2 weeks after surgery

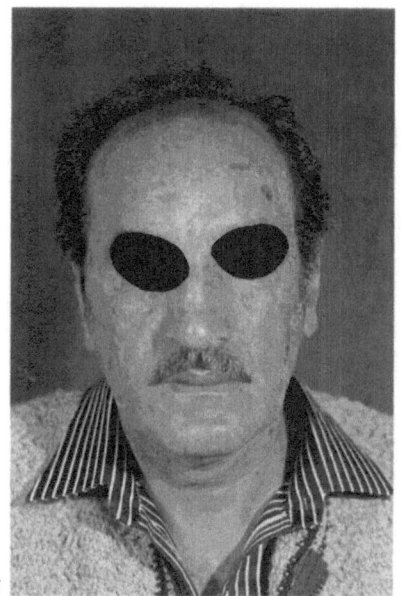

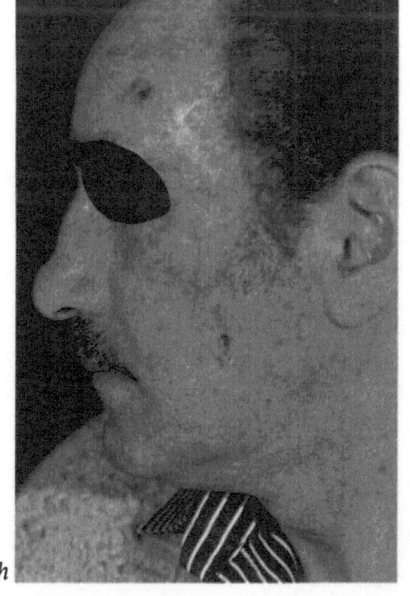

Figure 112

(k) — (m) The appearance 1 month after surgery

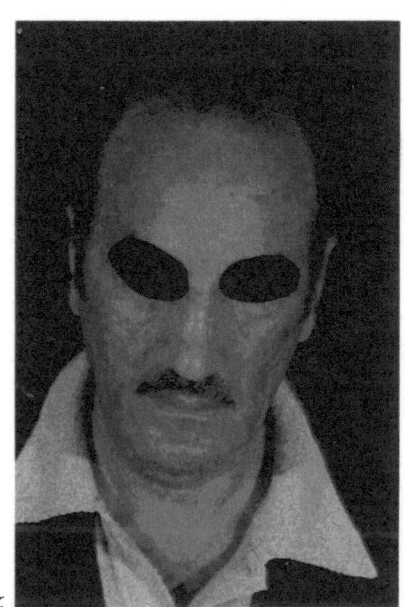

k

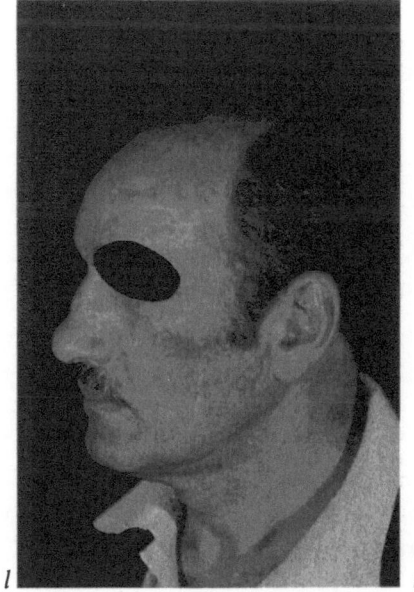

l

m

Figure 113 a–e

(a) *A young girl with a large pigmented nevus on the upper back. The procedure was performed under general anesthesia. In this case, test areas were performed in order to study the effect of treatment*

(b) *During treatment*

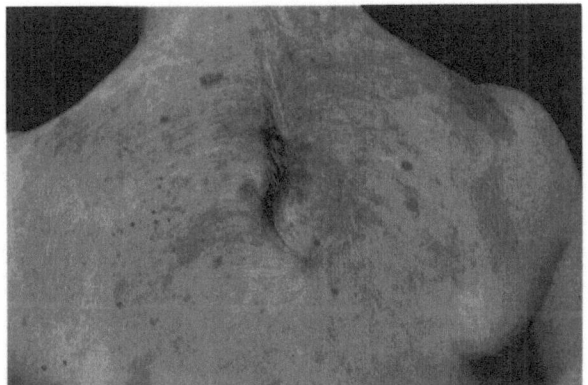

a

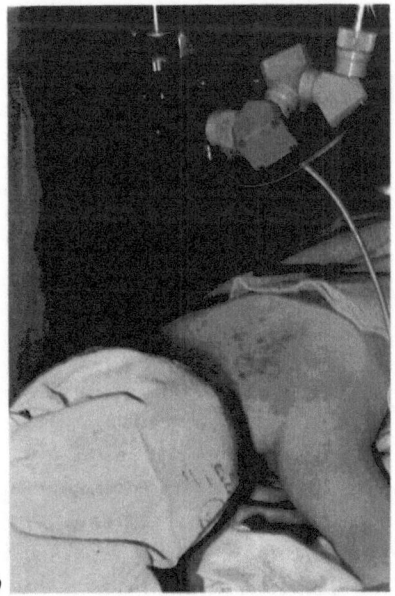

b

Figure 113

(c) The appearance 24 h after treatment
(d) The appearance after 2 weeks
(e) The appearance 1 month after treatment. It can be seen that pigment has disappeared from the areas treated

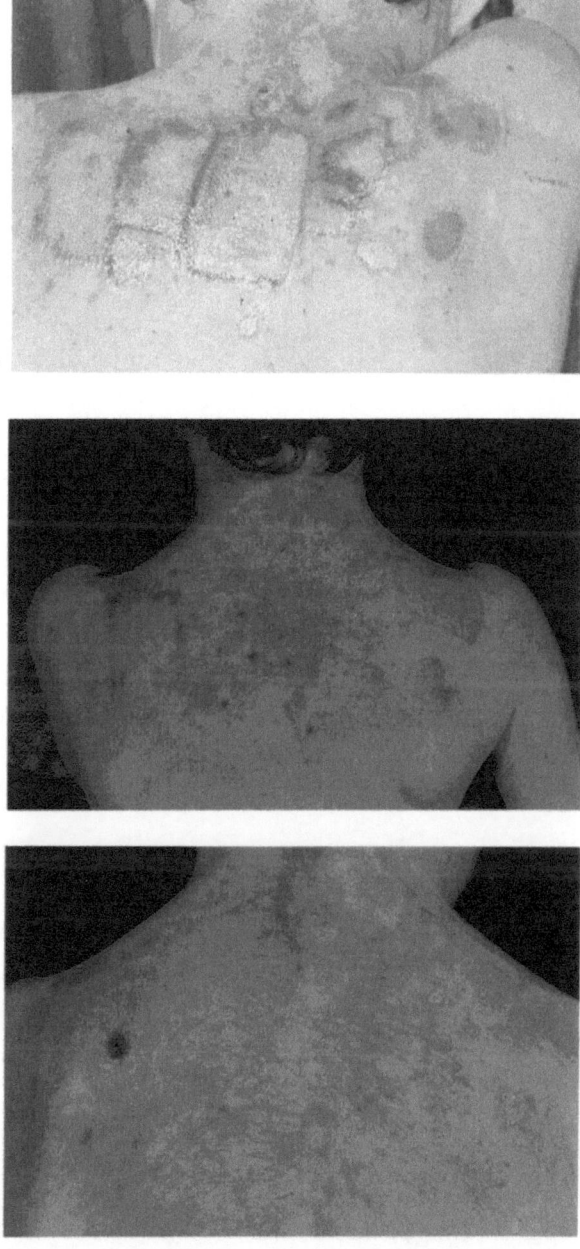

c

d

e

Figure 114a–d

*(a), (b) A patient with rhinophyma. The treatment was performed under local anesthe-
sia*
(c)   The appearance 1 week after treatment
(d)   The appearance 2 weeks after treatment

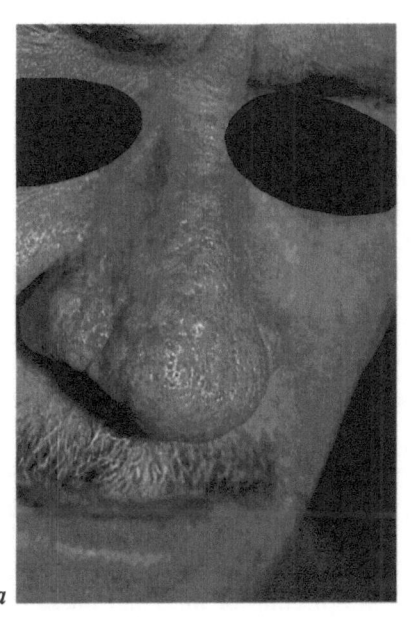

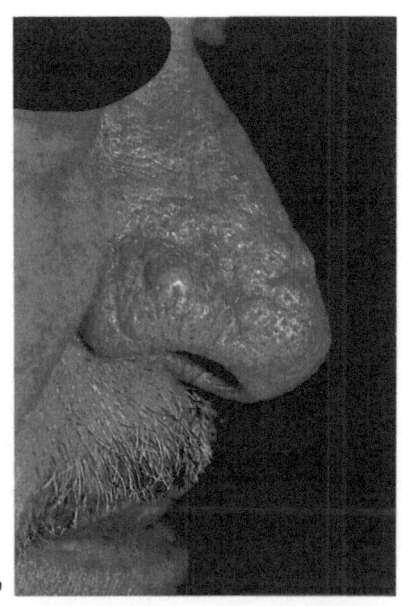

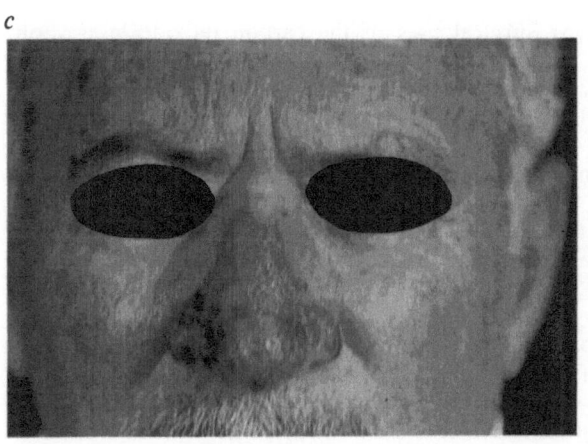

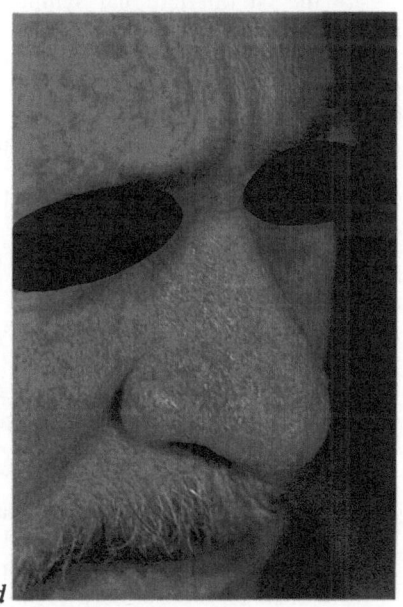

Publications on Laser Surgery

1. Kaplan I, Ger R (1973) The carbon dioxide laser in clinical surgery – a preliminary report. Isr J Med Sci 9:1:79–83
2. Kaplan I, Ger R (1973) Partial mastectomy and mammaplasty performed with CO_2 surgical laser – a comparative report. Br J Plast Surg 26:189–190
3. Kaplan I, Goldman J, Ger R (1973) Treatment of erosion of the uterine cervix by means of the CO_2 laser. Obstet Gynecol 41:795
4. Kaplan I, Ger R, Sharon V (1973) The CO_2 laser in plastic surgery. Br J Plast Surg 26:359–362
5. Frishman A, Gassner S, Kaplan I, Ger R (1974) Excision of subcutaneous fibrosarcoma in mice – a comparative experimental study of various methods. Isr J Med Sci 10:6:637–641
6. Kaplan I, Pariente R, Baiani G, Pochini M, Tantillo B, Toso M (1974) La resezion atipica sperimental di fegato mediante Laser al Biossido di Carbonio a flusso continuo IL Policlinico – Sez Chirurgica 81:5
7. Kaplan I, Gassner S, Schindel Y (1974) Carbon dioxide laser in head and neck surgery. Am J Surg 128:543–544
8. Ben-Bassat M, Gassner S, Kott I, Lavi E, Kaplan I (1975) A comparison between the scalpel and the CO_2 laser beam in the healing of intestinal anastomosis. Proceedings of the 1st International Symposium on Laser Surgery, Dec 1975. Jerusalem Academic Press, Jerusalem
9. Kaplan I (1975) CO_2 laser surgery. Harefuah 89:6:243–4
10. Kaplan I, Peled I (1975) The CO_2 laser in the treatment of superficial telangiectases Br J Plast Surg 28:214–215
11. Kaplan I, Peled I (1975) The carbon dioxide laser in plastic surgery. Rev Iberoam Cir Plas 1:4:35–45
12. Peled I, Kaplan I, Mattos S (1975) Surgical uses of the carbon dioxide laser. Folia Med 70:5
13. Kaplan I (ed) (1976) Laser Surgery I. Jerusalem Academic Press, Jerusalem
14. Ben-Bassat M, Kaplan I, Schindel Y, Edlan A (1976) The CO_2 laser in surgery of the tongue. Laser Surg 1:73–79
15. Labandter H, Kaplan I (1976) The treatment of haemangiomata using the CO_2 laser. Laser Surg 1:109–111
16. Peled I, Shohat B, Gassner S, Kaplan I (1976) Excision of subcutaneous Lewis lung carcinoma in mice – a comparative experiment. Laser Surg 1:66–69
17. Morein G, Kaplan I, Gassner S (1976) Laser-induced epiphysiodesis. Laser Surg 1:145–148
18. Ben-Bassat M, Ben-Bassat M, Kaplan I (1976) Electron microscopic studies of soft tissue incision by means of CO_2 laser. Laser Surg 1:95–100
19. Ben-Bassat M, Gassner S, Kaplan I, Kott I (1976) The healing process in experimental bowel surgery. Laser Surg 1:84–86

20. Kott I, Gassner S, Ben-Bassat M, Kaplan I (1976) The surgical knife and the CO_2 laser beam. Am J Proctol Gastroenterol Colon Rectal Surg 27:2:31
21. Labandter H, Kaplan I (1976) Onychogryphoses treated with the CO_2 surgical laser. Br J Plast Surg 29:102−1−3
22. Kaplan I, Sharon U (1976) Current laser surgery. Ann N Y Acad Sci 267:247−253
23. Peled I, Shohat B, Gassner S, Kaplan I (1976) Excision of epithelial tumors. CO_2 laser vs. conventional methods. Cancer Lett 2:41−46
24. Ben-Bassat M, Ben-Bassat M, Kaplan I (1976) A study of the ultrastructural features of the cut margin of skin and mucous membrane specimens excised by carbon dioxide laser. J Surg Res 21:77−84
25. Taube E, Glass I, Motovitz A, Kaplan I (1977) The CO_2 laser in veterinary surgery. Isr Vet Med Assoc 34:1:35
26. Meiraz D, Peled I, Gassner S, Ben-Bassat M, Kaplan I (1977) The use of the CO_2 laser for partial nephrectomy. Invest Urol 15:3:252−264
27. Kaplan I (1977) The Sharplan 791 CO_2 surgical laser in clinical surgery. In: Waidelich W (ed) Laser 77 Opto-Electronics. IPC Science and Technology Press
28. Labandter H, Kaplan I (1977) Experience with continuous laser in the treatment of suitable cutaneous conditions. J Dermatol Surg Oncol 3:527−530
29. Ben-Bassat M, Levy R, Kaplan I (1978) CO_2 laser in the treatment of Osler's disease. Br J Plast Surg 31:157−158
30. Kaplan I (1978) The Sharplan 791 Carbon dioxide surgical laser. Br J Clin Equip 227:229
31. Ben-Bassat M, Kaplan I, Levy R (1978) Treatment of hereditary haemorrhagic telangiectasia of the nasal mucosa with the carbon dioxide laser. Br J Plast Surg 31:157−158
32. Ben-Bassat M, Kaplan I, Schindel J, Edlan A (1978) The CO_2 laser in surgery of the tongue. Br J Plast Surg 31:155−156
33. Morein G, Gassner S, Kaplan I (1978) Bone growth alterations due to application of CO_2 laser beam to the epiphyseal growth plates − an experimental study of rabbits. Acta Orthop Scand 49:244−248
34. Kaplan I (ed) (1978) Laser Surgery II. Jerusalem Academic Press, Jerusalem
35. Kaplan I, Ascher PW (eds) (1980) Laser Surgery III. Jerusalem Academic Press, Jerusalem
36. Kaplan I (1980) The Sharplan CO_2 surgical laser in clinical surgery. Laser Surg 3:97
37. Taube E, Kaplan I, Glass I, Engelberg M (1980) Veterinary surgery by means of Sharplan CO_2 surgical laser. Laser Surg 3:98−99
38. Giler S, Ben-Bassat M, Kaplan I (1980) The use of the Sharplan CO_2 laser for lymph node dissection in cases of malignant melanoma. Laser Surg 3:100−106
39. Kaplan I (1980) The Sharplan CO_2 surgical laser in neonatal surgery. Laser Surg 3:197−200
40. Giler S, Ben-Bassat M, Taube E, Kaplan I (1980) The surgery of pilonidal sinus with the CO_2 laser. Laser Surg 3:201−203
41. Giler S, Gassner S, Ben-Uri R, Kaplan I (1980) The CO_2 laser in surgery of the pancreas − an experimental study. Laser Surg 3:211−216
42. Giler S, Ben-Bassat M, Gassner S, Kaplan I (1980) The CO_2 laser in surgery of the spleen − an experimental study. Laser Surg 3:217−233
43. Giler S, Kaplan I (1981) The use of the CO_2 laser for the treatment of cutaneous lesions in an outpatient clinic. Proceedings of 4th Congress of the International Society for Laser Surgery 1:1−4
44. Giler S, Kaplan I (1981) Multiple seborrheic keratoses treated with the CO_2 laser. Proceedings of 4th Congress of the International Society for Laser Surgery 1:5−8
45. Segal T, Nordenberg D, Giler S, Serebro I, Kaplan I (1981) The effect of the Sharplan CO_2 laser beam on dental structure. Proceedings of 4th. Congress of the International Society for Laser Surgery 12:25−28

46. Nordenberg D, Segal T, Giler S, Serebro I, Kaplan I (1981) The effect of the Sharplan CO_2 laser beam on dental caries: Sterilization a new approach. Proceedings of 4th Congress of the International Society for Laser Surgery 12:33−35
47. Giler S, Ben-Bassat M, Taube E, Kaplan I (1981) The use of the CO_2 laser in palliative surgery for cancer. Proceedings of 4th Congress of the International Society for Laser Surgery 23:12−15
48. Giler S, Kott I, Ben-Bassat M, Kaplan I (1981) The use of the CO_2 laser in anorectal surgery. Proceedings of 4th Congress of the International Society for Laser Surgery 23:36−38
49. Kott I, Reiss R, Giler S, Kaplan I (1981) The CO_2 laser in mastectomy: A comparative study. Proceedings of 4th Congress of the International Society for Laser Surgery 24:1−2
50. Tadir Y, Ovadia J, Zuckerman Z, Kaplan I (1981) Laparoscopic application of CO_2 laser. Proceedings of 4th Congress of the International Society for Laser Surgery 13:25
51. Kaplan I (1982) Current CO_2 laser surgery. Optics Laser Technol 41−42
52. Kaplan I (1982) The Sharplan CO_2 surgical laser in neonatal surgery. Ann Plast Surg 8:5:426−428
53. Kaplan I (1982) Current CO_2 laser surgery. Plast Reconstr Surg 69:3
54. Kaplan I, Sarig A, Ben-Bassat M (1982) Le laser à CO_2 en chirurgie pediatrique. L'Information Dentaire 3/6/82 – No. 22 Special Laser 2141−2143